SOUTH ASIAN INSTABILITY

This book examines key concerns in South Asian security through a fine-grained history of the accidental firing of a missile and its aftermath between two nuclear-armed states—India and Pakistan—with tense relations in March 2022. It consolidates the official statements, media discourse and debates within the strategic communities in both countries into a coherent narrative. It looks at the role of key institutions in the crisis such as the Indian Air Force (IAF), The Directorate of Air Staff Inspection-IAF, Indian Ministry of Defence (MOD), Pakistan Air Force (PAF), Pakistani Foreign Office (FO), Inter-Services Public Relations (ISPR) and others. The book also examines the missile accident's coverage in the international media and discussions in the global think-tank community.

Drawing on a host of resources, including published interviews of government officials, analyses in media, and strategic communities in India, Pakistan and the United States, this volume will be key reading for scholars and researchers of military and strategic studies, politics and international relations, public policy and South Asian studies.

Chaitanya Ravi is Assistant Professor and Chair of Public Policy at FLAME University, Pune, India. His research interests are in energy policy, environmental policy, public policy, international relations and aging. He is the author of *A Debate to Remember—The US-India Nuclear Deal* (2018) and his work has appeared in journals such as *Politics and Policy* and *Journal of Applied Gerontology*. He has taught energy policy, environmental policy, biology and microbiology at George Mason University, Virginia, and environmental studies at New York University. He holds a Masters in Environmental Science from Savitribai Phule Pune University, India, and a Bachelor of Science in Microbiology from D. Y. Patil Arts, Commerce and Science College, Pune, India.

South Asia in Context

South Asia in Context is a multidisciplinary series which offers a platform for groundbreaking research in new and established frontiers in South Asian Studies from scholars across the world. It informs on and spotlights the diverse culture, history, society, emerging economy, religion, and geography as well as politics of the various South Asian countries.

Bringing together contributions from across the humanities and the social sciences, the Series aims to provide a critical and in-depth analysis of the many histories, current affairs, and possible futures as well as challenges faced by the region. The books in the series are grounded in extensive theoretical and empirical research, covering both micro and macro themes of global and local relevance. They contribute to the expanding oeuvre of contemporary South Asian Studies and bring into focus hitherto unexplored areas of research.

Books in the series are published simultaneously in UK/ US and South Asia editions, as well as in e-book format. Please contact rioeditorial@tandfindia.com if you wish to submit a new proposal.

Indigenous Question, Land Appropriation, and Development
Understanding the conflict in Jharkhand, India
Gautam Pingali

African Clusters in India
Koyal Verma

For more information about this series, please visit: https://www.routledge.com/South-Asia-in-Context/book-series/SAIC

SOUTH ASIAN INSTABILITY

The India-Pakistan Missile Accident

Chaitanya Ravi

Routledge
Taylor & Francis Group
LONDON AND NEW YORK

Designed cover image: Getty Images

First published 2025
by Routledge
4 Park Square, Milton Park, Abingdon, Oxon OX14 4RN

and by Routledge
605 Third Avenue, New York, NY 10158

Routledge is an imprint of the Taylor & Francis Group, an informa business

British Library Cataloguing-in-Publication Data
A catalogue record for this book is available from the British Library

ISBN: 978-1-032-43766-8 (hbk)
ISBN: 978-1-032-89651-9 (pbk)
ISBN: 978-1-003-54390-9 (ebk)

DOI: 10.4324/9781003543909

Typeset in Sabon
by Deanta Global Publishing Services, Chennai, India

CONTENTS

ACKNOWLEDGMENTS

I thank FLAME University for the stimulating intellectual environment and administrative support which provided the enabling conditions for the completion of this book. I am also indebted to Routledge's commissioning, editing and production teams for their prompt correspondence, timely advice and quick processing of this book. I am grateful to my parents for their words of encouragement and moral support. My wife and my brother also deserve special thanks for their help in picking out the book cover and for their thoughtful comments on the manuscript.

INTRODUCTION

Accidental launches of cruise or ballistic missiles, even if unarmed, are a serious regional security risk in the South Asian region given their potential for escalation in the volatile relationship between nuclear-armed India and Pakistan. Such launches, especially during times of heightened tension such as in the aftermath of a terrorist attack or a surgical strike, risk further aggravating the dangerous mix of antagonistic nationalisms, rival territorial claims over the state of Kashmir, history of four wars, lack of high-level communication and deep mistrust thereby reducing the threshold for conflict. And yet, such accidental launches seem unavoidable in a tense South Asian security environment characterized by growing sophistication of delivery systems, regular testing of such systems and aggressive deployment postures.

On 10 March 2022, as the world was pre-occupied with Russia's invasion of Ukraine, Director General of Pakistan's Inter-Services Public Relations (ISPR), Major General Babar Iftikhar held a press briefing in which he made a startling claim. An Indian "supersonic flying object, most probably a missile"[1] which was unarmed had violated Pakistani airspace on the evening of 9th March, crashing near the Pakistani town of Mian Channu,[2] 124 km inside Pakistan.[3] This was the first instance in history of a nuclear-armed state accidentally launching a conventional, but also nuclear-capable missile,[4] deep inside the territory of another nuclear-armed state[5] with which it had fought four wars and had disputed borders.

A day later, on 11 March, the Indian Ministry of Defence (MOD) confirmed that a "technical malfunction" during routine maintenance had led to the "accidental firing of a missile."[6] The Ministry ordered a high-level Court of Enquiry and termed the incident as "deeply regrettable."[7] Although

DOI: 10.4324/9781003543909-1

Defence Minister Rajnath Singh did not identify the errant missile in his statement to Parliament, defense officials reckoned that it was a BrahMos supersonic cruise missile.[8] The Indian government's narrative on the cause of the accident changed from the initially identified technical malfunction to human error involving a deviation from standard operating procedures.[9] A high-level Court of Enquiry announced by the Indian government held three Indian Air Force (IAF) officers—a group captain, a wing commander and a squadron leader—responsible for a deviation from standard operating procedures and dismissed them from service.[10] Media reports have claimed that Pakistan was prepared to retaliate to the errant Indian missile but chose to hold back.[11] The claims of Pakistan's aborted retaliatory strike have implications for South Asian stability.

Particularly worrying in this regard is the Pakistani elite's refusal to accept India's explanation of the missile accident as a human error and the results of India's Court of Enquiry which dismissed three IAF officers for deviation from Standard Operating Procedures (SOPs). The senior Pakistani leadership including National Command Authority adviser Lt Gen (retired) Khalid Ahmed Kidwai,[12] have indicated that the Indian launch was a deliberate act of a country spurred by an aggressive religious nationalist ideology.[13] The belief that India deliberately fired a missile into Pakistan is also shared by Pakistani academics,[14] strategists,[15] military[16] and far-right commentators with Indian motives behind the launch being ascribed as evaluating "Pakistan's entire nuclear doctrine," to check its "response and reaction time" and to validate India's new "first-attack doctrine."[17] The lack of trust and timely communication risks creating a dangerous dynamic in which even an unintentional missile misfire incident in the future may be ascribed devious motives by the aggrieved side to the one from whose territory the missile was launched. More specifically, accidental missile launches can also lead to miscalculation due to the inability of the country at the receiving end to distinguish whether the missile system is armed or not and whether the warhead is conventional or nuclear in a region where there is already a fear of such "entanglement" of conventional and nuclear systems.[18]

Through sheer good fortune, the March 2022 missile misfire did not cause any casualties except for some minor damage to civilian property.[19] However, the threshold for conflict escalation could be lowered in case an accidental launch causes casualties and occurs during periods of tension in the overall relationship, such as during the Kargil War of 1999, after the 2008 Mumbai terrorist attack, during the 2016 Surgical Strike or the 2019 Balakot crisis. In addition to its serious implications for stability in India-Pakistan relations, the missile accident also raises questions about similar risks to stability in Sino-Indian relations if an errant missile from India were to violate Chinese airspace or vice versa.[20]

Further, geographic proximity imposes severe time constraints with the political leadership having to decide on whether to retaliate or not and on proportional retaliation, within a few minutes of the missile's launch. More complications arise due to the potential of an accidental missile launch to feed into the domestic political milieu of both countries (especially if it causes casualties or damages a sensitive military installation), impelling leaders on both sides to adopt nationalist-maximalist positions and constraining their room for behind-the-scenes diplomatic solutions.

Overall, the missile accident of March 2022 elicited restrained reactions from Pakistan and India and occurred in the backdrop of a tenuous peace resulting from a ceasefire along the Line of Control (LOC) and the International Border (IB) since February 2021.[21] This combination of maturity and an enabling environment may not exist each time such an accident occurs in the South Asian region with both India and Pakistan regularly testing missiles and with both having had failed missile tests with some involving the missiles veering off their intended course.

While most Indian ballistic and cruise missiles are tested off the coast of the eastern state of Odisha,[22] test firings of land-attack cruise missiles have been conducted overland at the Pokhran field firing range in the western state of Rajasthan bordering Pakistan.[23] Pakistan's failed cruise missile of 2020 was conducted from the Sonmiani test range in Balochistan.[24]

Both India and Pakistan have experienced cruise missile test failures. Indian failures include the failure of the third test of the indigenous technology cruise missile (ITCM) in 2022,[25] the extended range version of the BrahMos cruise missile in 2021,[26] the first test of the ITCM in 2020,[27] [28] the multiple test failures of the Nirbhay land-attack cruise missile (LACMs),[29] and the failure of the extended range version of BrahMos cruise missile 2009[30] etc. Known Pakistani failed cruise missile tests include the Babur 2 ground-launched cruise missile in 2020 and 2018.[31] With cruise missile tests comprising 19% of all Indian missile tests since the first documented test of the BrahMos in 2001 and 30% of all Pakistani missile tests since its first known cruise missile test in 2005,[32] the possibility of recurrence of an accidental launch of a land-attack cruise missile similar to the missile misfire of 2022 cannot be entirely ruled out.

There exists a considerable documentary record on the missile accident. Media sources in both countries including veteran journalists, editors and editorial boards have closely covered the immediate aftermath of the incident and the reactions of the government in both countries. The various statements given by top political leaders from both countries are also available. The Indian and Pakistani media and think-tanks also carried in-depth analyses of the incident by veteran strategists in both countries for South Asian and global security. The international media, including agenda-setting newspapers, also covered the incident, albeit to a limited extent. And finally, international strategic analysts, especially in the US, have also conducted in-depth analyses of the causes of the incident.

The aforementioned existence of the documentary record notwithstanding, no scholarly, book-length work has conducted an in-depth examination of the initial reactions of the Indian and Pakistani governments to the crisis, the role of key state institutions in both countries, the narratives advanced by the two states and their evolution over time and the major themes covered by the agenda-setting media in both countries. There is also no systematic consolidation and analysis of the major themes in the global media discourse on the accident, publicly available information on the Indian government's investigation into the accident, discourse of the strategic communities of both countries and the discourse of the global strategic community.

This book is a fine-grained history of the missile accident that consolidates the discourse around the accident into a coherent narrative by drawing from government statements, national and global media coverage, debates within the strategic communities in both countries and discussions among international experts. It also looks at the role of key institutions such as the Indian Air Force (IAF), Ministry of Defence (MOD), BrahMos Aerospace and others on the Indian side and the Pakistan Air Force (PAF), Inter-Services-Public Relations (ISPR), the Pakistani Foreign Office (FO) and others on the Pakistani side.

The theoretical basis of the book is its constructivist approach which delves into the Indian and Pakistani sides of the debate in order to understand the role of government institutions, the media and strategic communities, often in their own words through carefully edited quotes.[33]

Constructivism[34] is a major paradigm in international relations theory which holds that the behavior of states in the international system is "shaped by elite beliefs, collective norms and social identities."[35] Constructivists view the core focus of their approach as a preoccupation with the "social construction of international politics" resulting from their belief that the behavior of actors in international relations such as the interests they profess to hold or the structures within which they are constrained to operate "are defined by social norms and ideas rather than by objective or material conditions."[36] Constructivists posit that "the structures of human association are determined primarily by shared ideas rather than material forces" and that the "identities and interests of purposive actors are constructed by these shared ideas rather than given by nature."[37] Constructivists also reckon that "fundamental changes in international politics occur when beliefs and identities of domestic actors are altered thereby also altering the rules and norms that are constitutive of their political practices."[38]

Thus, contructivists tend to focus on the "impact of ideas" and consider the "interests and identities of states as a highly malleable product of specific historical processes."[39] This leads them to focus granular attention on the "prevailing discourse(s) in society because discourse reflects and shapes beliefs and interests, and establishes accepted norms of behavior."[40] An understanding of how "identities are constructed, what norms and practices accompany their production, and how they each construct each other" is an important preoccupation of the constructivist research program.[41]

The constructivist approach of closely examining the internal decision-making processes of a state through a fine-grained analysis of its elite discourse starkly contrasts the security-centered, realist approach which assumes an anarchic international system in which self-interested, "black-box" states are primarily concerned with survival and on the acquisition of military and economic power required for the same.[42] Constructivism also differs with liberalism's state-centered analysis of state behavior as resulting from the concerns such as economic development and belief in liberal values thus being amenable to reducing inter-state "conflictive tendencies"[43] through international institutions, economic co-operation and spread of democracy.[44]

The constructivist approach has been fruitfully deployed in the South Asian context to yield revealing insights on the dynamics of interactions between South Asian states. In "The Roots of Rhetoric," Haider K. Nizamani deploys the discourse analysis methodology to closely analyze the elite nuclear discourses of India and Pakistan by delving into their respective domestic contexts. He contends that "nuclear politics cannot be divorced from the issue of identity in both countries."[45] Nizamani uses the notion of "epistemic communities" to identify elites in India and Pakistan that constitute "the core of the nukespeakers."[46] Nizamani reckons that India's current nuclear discourse has undergone a major shift by "totally abandoning the Nehruvian ideal" of an India vigorously pursuing nuclear power while remaining resolutely opposed to the making of nuclear weapons.[47] Its current nuclear discourse constructs India as a deserving great power not given its due in the current international order and nuclear weapons are viewed as "one of the important means" to elevate the country's status and to deter adversary states.[48]

In contrast, Pakistan's nuclear discourse, which has remained constant since the times of former Prime Minister Zulfikar Ali Bhutto since the late 1960s, constructs Pakistan as a Muslim nation facing a conventionally stronger, perpetually hostile and nuclear-armed, Hindu India where nuclear weapons serve as deterrent and as a "harbinger of the Islamic bomb."[49] Nizamani concludes that "the interaction of the subcontinent with nukes amply testifies that these are political weapons" that have been put to multiple uses "ranging from uplifting the national ego to ensuring self-respect."[50]

In "Constructivism and Indian Foreign Policy," Priya Chacko deploys the constructivist approach to understand India's "approach to Pakistan, Kashmir and cross-border terrorism under the Narendra Modi regime," by developing a critical constructivist approach to understand change and continuity in Indian foreign policy.[51] She draws from the economic sociology literature from Bob Joseph's cultural political economy approach.[52] Analyzing the discourse of the Congress-led United Progressive Alliance (UPA) Government, Chacko points to the construction of a "transformational" state identity which defined India in terms of its "civilizational exceptionalism," its undemocratic neighbors, British colonialism, neoliberal reforms of the 1990s and the "rising power" discourse in the international media in the mid-2000s.[53] In this altered transformational identity, historic "signifiers" such as self-reliance which comprised the core of India's quasi-socialist economy were redefined to mean economic independence through liberalization, resulting in a growth-focused and regional peace-oriented foreign policy including one of sustained engagement and restraint towards Pakistan.[54] Chacko then chronicles the rise of an "aspirational identity" under the Bharatiya Janata Party(BJP), which while continuing the growth-focused and neighborhood-focused thrust of policies engendered by the UPA's transformational state identity, has also taken on additional signifiers such as "aspiring power" and "leading power."[55] She explains the BJP government's decision to conduct the surgical strikes against Pakistan as it attempts to unite the RSS (and its Hindu constituency of small farmers, small traders and small businesses) which was opposed big-bang reforms that favor large corporates, and "neo-middle class" in which urban Hindus have benefited from such reforms.[56]

In keeping with the constructivist approach's core precepts, the main assumption underlying this approach is that a chronological consolidation of the diverse issues, positions, actors and institutions on the Indian and Pakistani sides can provide a granular, social constructivist understanding of how two major South Asian nuclear weapon states managed a missile accident that could have potentially escalated into a more serious security crisis. The aforementioned constructivist approach is also extended to consolidate the global media discourse and discussions within the global strategic community.

The book's methodological focus involves a careful review and chronicling of the domestic discourse in India and Pakistan consisting of government statements, think-tank reports, think-tank discussions, print media, electronic media and social media from 9 March 2022 when the Indian missile was accidentally launched to 23 August 2022 when some findings of India's Court of Enquiry were released to the media. A secondary search from the period between 23 August 2022 and May 2023 of the aforementioned sources is also conducted to chronicle the litigation in the Delhi High Court in India pertaining to the missile accident.

A fine-grained history of the missile misfire attempted in this book uncovers a number of interesting insights. A reconstruction of the discourse in

India surrounding the sequence of events that led to the accidental launch (based on information in the public domain) consolidates media reports on the specific allegations made by India's Court of Enquiry against the erring IAF wing commander blamed for the accident. An example being that he failed to "perform external checks of the combat connectors of combat missiles, which continued to remain connected to FCS Junction Box 1&3."[57] An in-depth analysis of the accident also sheds some light on the steps claimed to have been taken by Indian authorities in the immediate aftermath of the accidental launch such as their decision to immediately cut-off the power to the vehicle so that remaining two missiles in the missile launcher would also not get launched.[58]

A sustained examination of the missile misfire also consolidates the available information regarding the dynamics of crisis communication (or lack thereof) between India and Pakistan with reliable media sources and strategic experts in India making contradictory claims regarding India using or not using the Director General of Military Operations (DGMO) military hotline to inform Pakistan. Insights also emerge regarding the supposed "hierarchy" of personnel within the IAF wherein some missile units are claimed to be manned by fighter pilots who have been "grounded" due to medical, competency and other reasons raising questions regarding the skill-sets and temperament of the personnel manning the missile launch units.[59]

The deployment of the constructivist approach also yields deeper insights regarding the connection between sophisticated weapons platforms and state identity. On the Indian side, the accident was understood as undermining India's self-perceived identity (bolstered by international discourse) as a responsible state in world affairs. Such a framing of the accident resulted in across-the-board calls from large sections of media and strategic communities for a swift and transparent inquiry and for severe punishment to the guilty officers. The brisk pace of the inquiry which dismissed three errant officers within six months of its announcement also indicated the Indian government's own commitment to actions that would minimize damage to its responsible state identity, which was one of the bases for India's nuclear de-hyphenation from Pakistan and for preferential treatment by the West.

On the Pakistani side, the accident reinforced the Pakistani identity as a Muslim state that would forever be at the receiving end of Hindu India's myriad conspiracies and as a garrison state which would have to resist Indian hegemony. The explicit link drawn by some senior Pakistani government officials, media commentators and strategic analysts between the missile intrusion and deliberate motives by the Hindutva-ideology influenced Indian government and the missile intrusion opened the space for conversations on preparation by Pakistan for a range of proportionate responses in case of a similar future intrusion. Alongside the aforementioned identity also existed a framing of India as an incompetent state with flawed standard operating procedures, faulty technology and errant personnel. This understanding sought to puncture the global understanding of India as a responsible state and the 'rising India' narrative in order to open the space for the re-hyphenation of India and Pakistan in South Asia.

Outline of this Book

Chapter 1 covers how the Indian and Pakistani governments responded in the wake of the crisis. A detailed, chronological discourse analysis is undertaken of the official statements by the Pakistani and Indian governments in the first few days of the crisis. The chapter also focuses on the key state institutions involved in the crisis such as the Indian Air Force (IAF), Directorate of Air Staff Inspection-IAF, Ministry of Defence (MOD), Pakistan Air Force (PAF), Pakistani Foreign Office (FO), Inter-Services Public Relations (ISPR) etc. Publicly available information about the evolution of the IAF probe and changes made by the IAF in Standard Operating Procedures (SOPs) for "storing missiles, launchers and the inspection process"[60] are also analyzed.

Chapter 2 looks at the role played by the Indian and Pakistani media in reporting on the missile crisis and analyzing its implications for regional security by focusing on English-language, agenda-influencing TV channels, newspapers, prominent digital outlets and social media snippets. The first part of the chapter focuses on the Indian media and the second part scrutinizes the Pakistani media. The chapter includes opinions as well as editorial positions of prominent newspapers in both countries. The chapter also chronicles anonymous statements from highly placed government sources to various media outlets in India.

Chapter 3 consolidates the debate and discussion on the missile accident within the strategic communities in India and Pakistan consisting of retired government bureaucrats, retired military officials, academics, think-tank experts and others. The chapter captures the convergences and divergences within the Indian strategic community regarding the missile accident. The chapter also consolidates the key themes within the discourse of the Pakistani strategic community on the missile accident.

Chapter 4 analyzes the manner in which the global media covered the accident and its attempts to distill the implications of the crisis for South Asian and global security. The chapter focuses on specific aspects such as the surprise in sections of the global media at the restrained manner in which India and Pakistan de-escalated the crisis.

Chapter 5 brings together the various analyses of the missile misfire by strategists in the West and around the world. It focuses on the nexus of technical, organisational, political and strategic issues highlighted by these strategists and their recommendations for crisis communication, de-escalation and prevention of future accidental launches.

Finally, the conclusion consolidates the key insights from the chapters and an in-depth discussion of the multi-level implications of the missile accident for South Asian security.

Overall, a constructivist analysis of the domestic discourse in India and Pakistan (and of the discourse at the international level) holds implications for state policies on a range of issues including crisis communication (use of military hotline) in the immediate aftermath of the accident, unambiguous communication to the media and strategic communities about the use or eschewal of the military hotline, serious bilateral discussions on expansion of existing agreements or formulation of new ones to prevent reoccurrence of such accidents, continuous conversations at the Track 2 level at the political and military levels on doctrinal clarity, prevention of entanglement of conventional and strategic systems, and implications of disruptive technologies such as UAVs and AI. Additional important implications for state policy also include serious intra-state consultations on existing standard operating procedures concerning inspection and launch sequence, psychological and skillset considerations for missile unit personnel, chain of command protocols with respect to reporting of an accident to the political authority and the need for dedicated crisis management centers for de-escalation etc.

Notes

1 "'India to Explain What Happened in Mian Channu,' Says DG ISPR after Indian Projectile Falls in Pakistan," *Dawn.com*, March 10, 2022, accessed October 9, 2023, https://www.dawn.com/news/1679289. DG ISPR Press Conference – March 102022, accessed October 9, 2023, https://www.youtube.com/watch?v=10lj66eRujA
2 Matt Korda, "Flying under the Radar: A Missile Accident in South Asia," accessed October 9, 2023, https://fas.tghp.co.uk/publication/flying-under-the-radar-a-missile-accident-in-south-asia/
3 Dawn.com, "'India to Explain What Happened in Mian Channu,' Says DG ISPR after Indian Projectile Falls in Pakistan," *DAWN.COM*, March 10, 2022, accessed October 7, 2023, https://www.dawn.com/news/1679289
4 Rashme Sehgal, "'Even If Authorised, Missile Firing Makes No Sense,'" *Rediff*, March 13, 2022, accessed October 9, 2023, https://www.rediff.com/news/interview/dr-bharat-karnad-even-if-authorised-missile-firing-makes-no-sense/20220313.htm
5 Matt Korda, "Flying under the Radar: A Missile Accident in South Asia," accessed October 9, 2023, https://fas.tghp.co.uk/publication/flying-under-the-radar-a-missile-accident-in-south-asia/
6 Press Information Bureau, "Statement on Accidental Firing of Missile," March 11, 2022, accessed October 9, 2023, https://pib.gov.in/pib.gov.in/Pressreleaseshare.aspx?PRID=1805148
7 Press Information Bureau, "Statement on Accidental Firing of Missile," March 11, 2022, accessed October 9, 2023, https://pib.gov.in/pib.gov.in/Pressreleaseshare.aspx?PRID=1805148
8 Peri, D. "IAF Completes Inquiry into Accidental Missile Firing Incident," *The Hindu*, April 9, 2022, sec. India., accessed October 6, 2023, https://www

.thehindu.com/news/national/iaf-completes-inquiry-into-accidental-missile-firing-incident/article65306093.ece, Huma Siddhiqui, "India Says Missile Landed in Pakistan by Accident; Check Details," *Financial Express*, March 11, 2022, accessed October 4, 2023, https://www.financialexpress.com/business/defence-india-says-missile-landed-in-pakistan-by-accident-check-details-2458430/

9 Rahul Bedi, "Did Rajnath Singh Subtly Suggest 'Human Error' Led to Accidental Missile Firing?" *The Wire*, March 17, 2022, accessed October 9, 2023, https://thewire.in/security/india-accidental-missile-firing-human-error-rajnath-singh, Rahul Singh, "IAF Sacks Three Officers for BrahMos Misfire into Pakistan in March," *Hindustan Times*, August 23, 2022, accessed October 9, 2023, https://www.hindustantimes.com/india-news/iaf-sacks-three-officers-for-brahmos-misfire-into-pakistan-in-march-101661263805711.html

10 A. Khan, "Reassessing the BrahMos Missile That Landed in Pakistan," *The Diplomat*, September 29, 2022, accessed October 9, 2023, https://thediplomat.com/2022/09/reassessing-the-brahmos-missile-that-landed-in-pakistan/

11 R. S. Sen and F. Mangi, "Indian Missile Misfire Prompted Pakistan for Retaliatory Strike," *Bloomberg*. Republished in *Deccan Herald*, accessed October 9, 2023, https://www.deccanherald.com/india/indian-missile-misfire-prompted-pakistan-for-retaliatory-strike-1091827.html, "Pak Nearly Launched Retaliatory Strike after India's Accidental Missile Launch," *Livemint*, March 16, 2022, accessed October 1, 2023, https://www.livemint.com/news/world/pakistan-nearly-launched-retaliatory-strike-in-response-to-india-s-accidental-missile-launch-11647424134509.html

12 B. S. Syed, "Concern over 'Hindu Extremists' Controlling Indian N-Arsenal," *DAWN.COM*, September 27, 2022, accessed October 1, 2023, https://www.dawn.com/news/1712196

13 A. Khan, "Reassessing the BrahMos Missile That Landed in Pakistan," *The Diplomat*, September 29 2022, *a*ccessed October 9, 2023, https://thediplomat.com/2022/09/reassessing-the-brahmos-missile-that-landed-in-pakistan/

14 "India Had Fired BrahMos into Pakistan under Ulterior Motives: Jaspal," *Pakistan Today*, September 28, 2022, accessed October 1, 2023, https://www.pakistantoday.com.pk/2022/09/28/india-had-fired-brahmos-into-pakistan-under-ulterior-motives-jaspal/

15 Dr Asma Shakir Khawaja, "Brahmos Fire on Pakistan: A Question Mark on Indian Credibility as Responsible State Actor," Global Village Space, March 30, 2022, accessed April 4, 2023, https://www.globalvillagespace.com/brahmos-fire-on-pakistan-a-question-mark-on-indian-credibility-as-responsible-state-actor/, Keynote Address by Lt. Gen. Khalid Ahmed Kidwai on 8th CISS-IISS International Workshop. *Center for International Strategic Studies*, September 21, 2022, accessed May 23 2023, https://www.youtube.com/watch?v=4WHbGmzE1tc

16 Hassan Aslam Shad, "Did India Purposely Fire a BrahMos Missile at Pakistan?" *The National Interest*, March 21, 2022, accessed May 23, 2023. https://nationalinterest.org/blog/buzz/did-india-purposely-fire-brahmos-missile-pakistan-201341

17 OpIndia Staff, "'India Planning to Invade, Yogi Celebrating Victory': Pakistani Experts Make Wild Claims about Alleged 'Indian Missile,'" *OpIndia* (blog), March 11, 2022, accessed October 3, 2023. https://www.opindia.com/2022/03/india-preparing-for-invasion-yogi-adityanath-celebrating-election-victory-pakistani-experts-wild-theories-about-alleged-indian-missile/

18 CSDR, "Skirting the Nuclear Threshold: Conflict De-Escalation and Strategic Stability in South Asia," *CSDR New Delhi*, March 31, 2022, accessed September 21, 2023, https://www.youtube.com/watch?v=65h66CFQv1w, Feroz Hassan

Khan, "Strategic Risk Management in Southern Asia," *Journal for Peace and Nuclear Disarmament* 5, no. 2 (August 18, 2022): 369–93, https://doi.org/10.1080/25751654.2022.2136878

19 Scroll Staff, "India's BrahMos Missile Misfire into Pakistan Not of Specific Concern, Says Nuclear Watchdog Chief," Text. *Scroll.in,* November 14, 2022, accessed October 3, 2023, https://scroll.in/latest/1037368/indias-brahmos-missile-misfire-into-pakistan-not-of-specific-concern-says-nuclear-watchdog-chief

20 D. Raghunandan, "The Curious Incident of the 'Misfired' Missile," *NewsClick*, March 17, 2022, accessed October 3, 2023, https://www.newsclick.in/curious-incident-misfired-missile

21 S. V. Krishna, "Bordering on Peace: Evaluating the Impact of the India-Pakistan Ceasefire," *Carnegie India*, February 24, 2022, accessed October 9, 2023, https://carnegieindia.org/2022/02/24/bordering-on-peace-evaluating-impact-of-india-pakistan-ceasefire-pub-86513

22 "India Successfully Carries out Night-Firing Test of Ballistic Missile in Odisha," *India Today*, June 27, 2019, accessed October 9, 2023, https://www.indiatoday.in/india/story/strategic-forces-command-night-firing-test-ballistic-missile-odisha-1557559-2019-06-27, "India Successfully Test-Fires New Version of BrahMos Supersonic Cruise Missile off Odisha Coast," *India TV* News Desk, January 20, 2022, accessed October 2, 2023, https://www.indiatvnews.com/news/india/india-successfully-test-fires-new-version-brahmos-supersonic-cruise-missile-off-odisha-coast-latest-news-updates-drdo-2022-01-20-755382, R. Srivastava, "DRDO-Developed Indigenous Technology Cruise Missile Successfully Test-Fired off Odisha Coast," *Zee News,* August 11, 2021, accessed October 9, 2023, https://zeenews.india.com/india/drdo-developed-indigenous-technology-cruise-missile-successfully-test-fired-off-odisha-coast-2383818.html

23 "Army Successfully Test-Fires BrahMos Missile," *The Times of India*, November 8, 2015, accessed September 28, 2023, https://timesofindia.indiatimes.com/city/jaipur/army-successfully-test-fires-brahmos-missile/articleshow/49707229.cms, V. Bhatia, "Indian Air Force Successfully Test-Fires Land-Attack Version of BrahMos Cruise Missile," *India Today*, May 27 2016, accessed September 28, 2023, https://www.indiatoday.in/india/rajasthan/story/indian-airforce-successfully-test-fires-land-attack-version-of-brahmos-cruise-missile-325859-2016-05-27

24 S. Gupta, "Pakistan's Effort to Launch 750km Range Missile Crashes," *Hindustan Times*, March 23, 2020, accessed September 28, 2023, https://www.hindustantimes.com/india-news/pakistan-s-effort-to-launch-750km-range-missile-crashes/story-UT5CbOR3K0uVojmiOYoKjO.html

25 H. K. Rout, "Homegrown Cruise Missile System Hit by Technical Snag," *The New Indian Express*, October 29 2022, accessed October 9, 2023, https://www.newindianexpress.com/states/odisha/2022/oct/29/homegrown-cruise-missile-system-hit-by-technical-snag-2512880.html

26 mint, "BrahMos Missile Fails during Testfiring, Falls Shortly after Takeoff," July 12, 2021, accessed September 27, 2023, https://www.livemint.com/news/india/brahmos-missile-fails-during-testfiring-falls-shortly-after-takeoff-11626111853159.html

27 H. K. Rout, "Homegrown Cruise Missile System Hit by Technical Snag," *The New Indian Express*, October 29, 2022, accessed October 9, 2023, https://www.newindianexpress.com/states/odisha/2022/oct/29/homegrown-cruise-missile-system-hit-by-technical-snag-2512880.html

28 V. Raghuvanshi, "India's Nirbhay Cruise Missile Test Fails," *Defense News,* October 12, 2020, accessed September 26, 2023, https://www.defensenews.com/global/asia-pacific/2020/10/12/indias-nirbhay-cruise-missile-test-fails/

29 Dawn.com, "Indian Cruise Missile Test Fails for the Fourth Time," *DAWN.COM,* December 21, 2016, accessed September 26, 2023, http://www.dawn.com/news/1303661
30 G. M. Radhakrishnan, "'BrahMos Failure Was Due to Software Glitch,'" *India Today*, accessed October 9, 2023, https://www.indiatoday.in/web-exclusive/story/brahmos-failure-was-due-to-software-glitch-39217-2009-02-09
31 S. Gupta, "Pakistan's Effort to Launch 750km Range Missile Crashes," *Hindustan Times*, March 23, 2020, accessed September 28, 2023, https://www.hindustantimes.com/india-news/pakistan-s-effort-to-launch-750km-range-missile-crashes/story-UT5CbOR3K0uVojmiOYoKjO.html, B. WION Sharma, "Pakistan's 'Babur II' Missile Suffers Second Consecutive Setback; Crashes after Two Minutes," *WION*, March 23 2020, accessed September 9, 2023, https://www.wionews.com/south-asia/pakistans-babur-ii-missile-suffers-second-consecutive-setback-crashes-after-two-minutes-288073, Staff, "Big Setback For Pakistan As Its Under-Development Babur-II Cruise Missile Crashes In Second Consecutive Test," *Swarajya*, March 23, 2020, accessed September 25, 2023, https://swarajyamag.com/insta/big-setback-for-pakistan-as-its-under-development-babur-ii-cruise-missile-crashes-in-second-consecutive-test-failure, M. H. Kristensen, Matt Korda, and Eliana Johns, "Pakistan Nuclear Weapons, 2023," *Bulletin of the Atomic Scientists,* September 11 2023, accessed October 9, 2023, https://thebulletin.org/premium/2023-09/pakistan-nuclear-weapons-2023/
32 S. Cotton and P. Anne, "Cruising for a Bruising: How Cruise Missiles Are Threatening Norms between India and Pakistan," *The Nuclear Threat Initiative*, November 15, 2019, accessed September 2, 2023, https://www.nti.org/analysis/articles/cruising-for-a-bruising-how-cruise-missiles-are-threatening-norms-between-india-and-pakistan/, I. Gilani, "ANALYSIS – Military-Level Contacts Essential to Avoid 'Accidental' War in South Asia," *Anadolu Agency,* March 20, 2022, accessed September 1, 2023, https://www.aa.com.tr/en/analysis/analysis-military-level-contacts-essential-to-avoid-accidental-war-in-south-asia/2540734
33 For a similar deployment of the constructivist approach to analyze the US-India nuclear deal, see Chaitanya Ravi, *A Debate to Remember: The US-India Nuclear Deal,* 1st ed. (Oxford University Press, 2018), https://doi.org/10.1093/oso/9780199481705.001.0001
34 The term was first used by Professor Nicholas Onuf. See N. G. Onuf, *World of Our Making: Rules and Rule in Social Theory and International Relations* (Columbia, SC: University of South Carolina Press, 1989).
35 S. M. Walt, "International Relations: One World, Many Theories," *Foreign Policy* 110, no. 29 (1998): 38, doi:10.2307/1149275
36 J. S. Barkin, "Realist Constructivism," *International Studies Review* 5, no. 3 (2003): 326, https://www.jstor.org/stable/3186573
37 A. Wendt, *Social Theory of International Politics* (Cambridge: Cambridge University Press, 1999), 1. A. Wendt, "Anarchy Is What States Make of It: The Social Construction of Power Politics," *International Organization* 46, no. 2 (1992): 391–425, https://doi.org/10.1017/S0020818300027764
38 R. Koslowski and F. V. Kratochwil, "Understanding Change in International Politics: The Soviet Empire's Demise and the International System," *International Organization* 48, no. 2 (1994): 215–47, https://doi.org/10.1017/S0020818300028174.
39 S. M. Walt, "International Relations: One World, Many Theories," *Foreign Policy* 110, no. 29 (Spring 1998): 40–41.
40 S. M. Walt, "International Relations: One World, Many Theories," *Foreign Policy* 110, no. 29 (Spring 1998): 40–41.

41 T. Hopf, "The Promise of Constructivism in International Relations Theory," *Intermational Security* 23, no. 1 (1998): 192.
42 S. M. Walt, "International Relations: One World, Many Theories," *Foreign Policy* 110, no. 29 (1998): 38, doi:10.2307/1149275.
43 M. S. Richard Little, *Perspective on World Politics 2005*, London. (Routledge, 2005), 388.https://www.taylorfrancis.com/books/mono/10.4324/9780203300527/perspectives-world-politics-richard-little-michael-smith
44 S. M. Walt, "International Relations: One World, Many Theories," *Foreign Policy* 110, no. 29 (1998): 38, doi:10.2307/1149275.
45 H. K. Nizamani, *The Roots of Rhetoric* (Westport, CT: London, 2000), Praeger. 13.
46 H. K. Nizamani, *The Roots of Rhetoric: Politics of Nuclear Weapons in India and Pakistan* (New Delhi: India Research Press, 2001), 15.
47 H. K. Nizamani, *The Roots of Rhetoric: Politics of Nuclear Weapons in India and Pakistan* (New Delhi: India Research Press, 2001), 143.
48 H. K. Nizamani, *The Roots of Rhetoric: Politics of Nuclear Weapons in India and Pakistan* (New Delhi: India Research Press, 2001), 144.
49 H. K. Nizamani, *The Roots of Rhetoric: Politics of Nuclear Weapons in India and Pakistan* (New Delhi: India Research Press, 2001), 143–44.
50 H. K. Nizamani, *The Roots of Rhetoric: Politics of Nuclear Weapons in India and Pakistan* (New Delhi: India Research Press, 2001), 148.
51 H. V. Pant, ed., *New Directions in India's Foreign Policy: Theory and Praxis* (Cambridge: Cambridge University Press, 2019), 49–55.
52 H. V. Pant, ed., *New Directions in India's Foreign Policy: Theory and Praxis* (Cambridge: Cambridge University Press, 2019), 55.
53 H. V. Pant, ed., *New Directions in India's Foreign Policy: Theory and Praxis* (Cambridge: Cambridge University Press, 2019), 56.
54 H. V. Pant, ed., *New Directions in India's Foreign Policy: Theory and Praxis* (Cambridge: Cambridge University Press, 2019), 57–58.
55 H. V. Pant, ed., *New Directions in India's Foreign Policy: Theory and Praxis* (Cambridge: Cambridge University Press, 2019), 59.
56 H. V. Pant, ed., *New Directions in India's Foreign Policy: Theory and Praxis* (Cambridge: Cambridge University Press, 2019), 60–61.
57 PTI, "HC Seeks Centre's Stand on Challenge to Dismissal of Air Force Officer over Brahmos Misfire," *ThePrint* (blog), March 14, 2023, accessed September 2, 2023, https://theprint.in/india/hc-seeks-centres-stand-on-challenge-to-dismissal-of-air-force-officer-over-brahmos-misfire/1443108/
58 "Indian BrahMos with 'a Mind of Its Own' Lands in Pakistan & CBMs to Prevent Unintended Escalation," *The Print*, March 16, 2022, accessed October 14, 2023. https://www.youtube.com/watch?v=c7aJrZdKZm8
59 K. P. Sanjeev Kumar, "BrahMos 'Misfire': Indian Govt's Dismissal of 3 Officers Raises Key Questions," *TheQuint*, August 25, 2022, accessed September 24, 2023, https://www.thequint.com/opinion/brahmos-misfire-indian-govts-dismissal-of-3-officers-raises-key-questions
60 A. Banerjee, "IAF Changes Sops on Missile Storage after Accidental Fire," *The Tribune*, April 11, 2022, accessed October 24, 2023, https://www.tribuneindia.com/news/nation/iaf-changes-sops-on-missile-storage-after-accidental-fire-385550

1

THE BIRTH OF A CRISIS

How the Two Governments Responded

On 11 March 2022, even as much of the world was pre-occupied with Russia's massive invasion of Ukraine and its far-reaching ramifications for European security, a missile accident with crucial implications for South Asian security occurred. It was a cross-border, unauthorized, accidental launch of a high-speed, unarmed, cruise missile from India across the international border deep into Pakistan, without any prior notification. That the missile did not result in loss of lives and caused only minor damage to civilian property in Mian Channu, Pakistan was the result of sheer good fortune. The BrahMos supersonic cruise missile, identified by anonymous government sources in the Indian press as the missile that was accidentally launched,[1] has been claimed by Indian press reports[2] and understood by Pakistan,[3] as being nuclear-capable (although the US National Air and Space Intelligence Centre still classifies it on the basis of publicly available evidence as a conventional missile).[4]

What if the unarmed but speeding missile had inadvertently landed in a densely populated urban area in Pakistan killing scores of people and causing extensive damage to property? And what if the accidental launch had occurred during periods of high tension and low trust in the adversarial India-Pakistan relationship such as the immediate aftermath of the Mumbai terror attacks, Pathankot or Pulwama? Given the *Bloomberg* report that Pakistan briefly considered but eventually held back on a retaliatory strike with a similar missile,[5] the implications of the accidental launch by India for South Asian security were indeed serious. An especially worrying question arises as to whether Pakistan was able to immediately recognize the intruding missile as unarmed or whether its decision to retaliate was influenced by genuine confusion over whether the missile was carrying a warhead or not.

DOI: 10.4324/9781003543909-2

A careful examination of the initial reactions of the Indian and Pakistani governments to the missile accident, the narratives advanced by both governments to justify their respective version of the sequence of events, the evolution of the Indian probe into the accident and the Pakistani reaction to its findings holds the potential to provide important insights regarding crisis-management and strategic communication capabilities in South Asia.

Chapter 1 undertakes a fine-grained examination of the documentary record to reconstruct the history of the first few days of the crisis, the initial reactions of the two governments, the evolution of official responses over the next few months and initial findings of the Indian Court of Inquiry. The chapter begins with a brief history of recent India-Pakistan relations in order to highlight the particular set of circumstances that resulted in the successful management of the crisis by the two governments.

A Rare Period of Stability in an Otherwise Adversarial Relationship

The usually tumultuous relationship between India and Pakistan experienced a rare period of calm in 2021, lasting through the early months of 2022. This was in contrast to the steady deterioration of ties in the previous years[6] with the 2016 terror attacks on the Pathankot Air Base and the terror attacks in the same year on the Indian Army brigade headquarters near Uri town. The Uri attack resulted in India's cancellation of its participation in the 19th SAARC summit[7] and retaliatory Indian cross-border "surgical strikes."[8]

Ties nosedived further with the 14 February 2019 terror attack on a military convoy killing 40 Indian soldiers in Pulwama[9] and retaliatory airstrikes by India on Balakot on 26 February, deep inside Pakistan on a claimed Jaish-e-Mohammed terrorist training camp.[10] India's decision to end Kashmir's special status on 5 August 2019 by abrogating Article 370 and its bifurcation of Kashmir into two centrally-administered union territories[11] further soured ties—with a furious Pakistan suspending trade and expelling the Indian High Commissioner.[12]

Ties began to stabilize after February 2021 following a joint statement of Indian and Pakistani Director General of Military Operations (DGMO)'s committing to "strict observance of all agreements, understandings and cease firing along the Line of Control and all other sectors."[13] A year-old ceasefire resulted with ceasefire violations decreasing sharply from 5,100 in 2020 to six in 2021.[14]

Another contributor to stabilizing India-Pakistan relations was Pakistan's announcement of its first-ever National Security Policy in January 2022. The policy emphasized peace with immediate neighbors, economic diplomacy and better trade ties with India even in the absence of a final resolution on Kashmir provided there was progress in bilateral talks.[15] The same month, India and Pakistan also exchanged their annual list of civilian prisoners and fishermen in each other's custody.[16] A statement supporting the resumption of India-Pakistan trade, suspended since Delhi's 2019 revocation of Kashmir's special status, was made in February 2022 by Abdul Razak Dawood, adviser to then Pakistan Prime Minister Imran Khan.[17] Both India and Pakistan abstained in March 2022 on a UN resolution condemning Russia's invasion of Ukraine.[18] The 117th meeting of the India-Pakistan Permanent Indus Commission took place in a cordial atmosphere from 1–3 March 2022 in Islamabad.[19]

The "High Speed Flying Object" and Pakistan's Cautious Initial Reaction

This brief period of calm in the India-Pakistan relationship would soon be interrupted. There was a brief, seemingly unrelated, report in the Pakistani press on 9 March claiming that a Pakistani "fighter training plane" had crashed near Mian Channu collapsing the "roofs of the houses near the crash-site."[20] Another report claimed the crash site as being near a toll plaza in Mian Channu.[21] Early on 10 March, the *Dawn* newspaper claimed that a "private aircraft," "likely to be used for training purposes crashed on a cold storage at Chak 115 in the middle of the fields" in Mian Channu, Khanewal district with its lone occupant ejecting safely.[22] It further reported that the police had prevented rescue personnel from coming close to the crash site as the army had launched a search operation to collect debris of the jet.[23] The confusion surrounding events was somewhat cleared on the evening of March 10 in a press conference by the Director General, Inter-Services Public Relations (DG-ISPR), Major Gen Babar Iftikhar.

Reading from a prepared statement, Iftikhar alleged that, on 9 March at 18:43 hours, "a high speed flying object was picked up inside Indian territory" by the Air Defense Operations Centre of Pakistan Air Force. He continued that the "object suddenly maneuvered towards Pakistani territory and violated Pakistan's airspace, ultimately falling near Mian Channu at 18:50hrs." Iftikhar expressed relief that there was no loss of human lives or injuries even as he assured that the Pakistan Air Force continuously tracked the "complete flight path of the flying object from its point of origin near Sirsa in India to its point of impact near Mian Channu" and initiated the "requisite tactical actions in accordance with the SOPs." He further worried that the flight path of the object endangered several domestic and international passenger flights both in India and Pakistan and also posed a threat to civilian lives and property at Mian Channu.

The Pakistani prepared statement did not take a clear position on what the flying object was and whether its violation of Pakistani airspace was inadvertent or deliberate. Iftikhar signed off by merely saying that the onus was on India to explain the cause of the "incident" and it "nevertheless shows their disregard for aviation safety and reflects very poorly on their technological prowess and procedural efficiency." Iftikhar ended with a strong protest on this "flagrant violation" of Pakistani airspace and warned against "recurrence of any such incident in the future."[24]

Overall, the Pakistani prepared statement was cautious in that it refused to identify the flying object as a missile, sought to assure that the Pakistan Air Force had continuously tracked it and attempted to use the incident to depict India as an irresponsible state possessing unreliable technology with a callous disregard for aviation safety. Called on by Iftikhar to explain the technical details of the incident, Air Vice Marshall Tariq Zia also began by referring to a "high speed object" being picked up by Pakistani air defense systems close to Sirsa, 104 km inside Indian territory, travelling in excess of Mach 2.5, at an altitude of around 40,000 feet. He informed that the initial course of the "projectile" was "south-westwardly," heading towards the Mahajan field-firing ranges but there was a right "track change" after travelling 70–80 km to head north west towards Pakistani airspace while "maintaining the same altitude and speed." The total distance travelled inside Pakistan was 124 km with the total time from initial pickup to the impact on ground being 6 minutes and 46 seconds.

Of this, the time in Pakistani territory was 3 minutes and 44 seconds with the Pakistani air defense system having a "solid and continuous tracking of this projectile." Zia assured that a "forensic analysis" was being conducted of the "object" and for the first time in the Pakistani briefing, speculated that "it was a supersonic missile, an unarmed missile and it was launched from the ground." In keeping with Pakistan's cautious stance, the map displayed during Zia's briefing avoided any categorical reference to the Indian object being a missile and was titled "Indian Projectile Profile" with the time of initial pickup (18:43:43 PST), border crossing (18:46:45 PST) and impact point (18:50:29 PST) mentioned.[25]

Probably a Supersonic Missile

In the Q&A session with journalists that followed the press conference, and in response to a question on whether the incident was deliberate or inadvertent, Iftikhar concurred with Zia's assessment that "it was a supersonic flying object, most probably a missile" and "it was certainly unarmed." He also assured that there was no communication from the Indian side via the military hotline and that there was no conversation between the DGMOs of both countries. In response to another question, he reiterated that the incident demonstrated India's lack of "professional prowess" of the "human resource teams" handling the missile and the lack of reliability of Indian technology.[26] He also informed that the missile had travelled a total of 260 km. Tariq Zia concluded the press conference by pointing out that the Indian "projectile" had followed a trajectory and an altitude that was close to two active passenger flight routes in India and Pakistan with the potential to threaten many international and domestic flights.[27]

On 10 March, the Pakistan Foreign Office (PFO) summoned the Indian Charge d'Affaires and registered a "strong protest" over the "unprovoked violation of its airspace by an Indian origin "super-sonic flying object" which entered into Pakistan from Suratgarh India at 18:43 hours (PST) on 9 March 2022 and fell to ground near Mian Channu city in Pakistan at around 18:50 hrs the same day."[28] The PFO also repeated the criticisms made by the DG-ISPR in his prepared statement of the flying object endangering domestic and international flights in Pakistani airspace and civilian lives on the ground.[29] Terming the incident as "irresponsible" the PFO demanded a "thorough and transparent investigation" with its findings to be shared with Pakistan.[30]

India's Initial Reaction: Regret, Technical Malfunction and a Court of Inquiry

On the Indian side, a 11 March article in *The Print* informed that the Indian defense establishment was looking into Pakistani airspace violation claims with an official statement likely to be released "by the evening."

> Sources said that the description so far matches that of the BrahMos supersonic missile and the fact that there was no warhead means it was a test firing. However, they remained mum on whether it was actually a BrahMos or not...Sources explained that the flight path taken by the BrahMos depends on the target....While Sirsa does not have a BrahMos base, the missile system is operated from specialised trucks and can be moved from one location to another. Such missile tests usually happen on the Eastern flank and from the Andaman and Nicobar Command.[31]

At 18:33 on 11 March, two days after the accident, the Indian Ministry of Defence issued a terse statement in which it confirmed that a "technical malfunction" during "routine maintenance" had led to the "accidental firing of a missile."[32] The statement assured that the Indian government had taken a "serious view" of the accident and had ordered a high-level Court of Inquiry.[33] The statement termed the "incident" as "deeply regrettable" while also expressing relief that no lives had been lost.[34]

The Indian statement, albeit brief, confirmed that a missile had been accidentally fired into Pakistan due to a claimed technical malfunction although it did not provide any details about the type of missile fired or whether it was indeed a supersonic missile (as Pakistan claimed). The statement also sought to reassure Pakistan and the international community that an institutional mechanism (Court of Inquiry) was being put in place to investigate the incident. The statement's expression of regret over the accident avoided placing the blame on the Indian military. And finally, the statement's expression of relief over no lives being lost seemed to be carefully worded to emphasize the non-seriousness of the accident.

Errant Missile as "Nuclear Capable" BrahMos

While the Indian defense ministry's official statement was scant on details regarding the type of missile launched, "an unnamed Pakistani official told *Reuters* on 12 March that the missile was a nuclear-capable, land-attack BrahMoscruise missile jointly developed by Russia and India."[35] Pakistani officials also claimed that India did not alert Pakistan through the military hotline immediately after the accidental launch.[36] In the absence of any communication from the Indian side, the Pakistan Air Defense Operations Centre

> immediately suspended all military and civilian aircraft for nearly six hours, and reportedly placed frontline bases and strike aircraft on high alert. Defence sources stated that these bases remained on alert until 13:00 PKT on March 14th.[37]

The BrahMos is a ramjet-powered, supersonic cruise missile developed as part of a joint venture in 1998 (BrahMos Aerospace)[38] between India's Defence Research and Development Organisation (DRDO) and Russia's NPO Mashinostroyenia and is capable of being launched from land, sea and air platforms with a travel-speed of approximately Mach 2.8 (3,450 kmph or 2,148 mph).[39] The name BrahMos represents the "fury of Brahmaputra and the grace of the Moskva Rivers" and is a medium-range missile.[40] Indian media reports have claimed the missile to be the "world's fastest cruise missile" with a maximum speed of Mach 4.[41] The missile's first-stage is a solid-fuel rocket which provides the initial acceleration while the second stage comprises of a liquid-fuelled ramjet engine which further boosts its speed to Mach 2.8 to 3 to hit targets within a strike range of 290 km, recently increased to 400 km.[42] As a "terrain hugging" missile the BrahMos can fly "circuitous paths at supersonic speeds" making the interception of the missile difficult.[43]

The missile is claimed to operate on a "fire and forget principle,"[44] with a pre-set target and an inertial or satellite GPS system which helps it navigate to the target.[45] While the US National Air and Space Intelligence Center (NASIC) claims that the BrahMos has a range of less than 300 km,[46] the Indian Ministry of Defence claimed on 20 January 2022 that the missile's range had been extended "by changing the missile's trajectory, modifying its control system and increasing indigenous content."[47] Sources in the Indian defense establishment have also claimed that the missile's range was extended due to "certain software changes that had to be made....with the help of Russians."[48] While the Defence Ministry did not officially confirm the new extended range, official sources informed that it was in excess of 500 km.[49] Although Indian press reports have claimed the BrahMos to be capable of carrying a nuclear warhead,[50] NASIC still categorizes it as conventional given the lack of public evidence of its capability to carry a nuclear warhead.[51] India is currently developing a hypersonic version of the missile (BrahMos II) capable of speeds of Mach 5 to Mach 7.[52]

More Details Emerge via Anonymous Leaks

On 12 March, a day after the Indian Defence Ministry's statement, more details emerged in the media "through sources in the defence and security establishment." The sources claimed that the missile that breached Pakistan's airspace was an "unarmed BrahMos supersonic missile, a practice version" that was "accidentally fired during an inspection at a secret satellite base of the Indian Air Force (IAF)" and not from Sirsa, Haryana as claimed by Pakistan.[53] The sources also informed that the missile did not suddenly veer off its flight path into Pakistan but followed a correct trajectory it would have in case of a conflict.[54] They explained the combination of safety mechanisms to prevent the accidental launch of a missile.

> Sources in the know expressed surprise over the "accidental launch", while noting that the missile system has several mechanical and software safety mechanisms to prevent any kind of accidental firing. They said that the missile has multiple geo-locations of targets fed into the system, which need to be selected or new ones added. Then there are multiple steps of clearance, including the entry of certain codes before the countdown starts, after which the missile goes into auto mode, sources said. Asked why there was no warhead, sources explained that it was a practice missile, which meant that everything was exactly the same as an armed BrahMos missile except for the warhead.[55]

The sources also claimed that India had informed the "relevant Pakistani authorities" regarding the accidental launch "long before" the DG-ISPR's press conference, although they "refused to divulge the specific time and at what level the Pakistanis were informed."[56] They further averred that the trajectory of the missile displayed by Pakistan was wrong, demonstrating its inability to either track the missile's flight path, mount an effective air defense (especially given its flight duration of 3 minutes and 44 seconds inside Pakistani territory) and accurately pinpoint its launch site.[57]

A 13 March article in the *Times of India* reported that Indian Air Force (IAF) had launched a probe into how "one of its units accidentally fired the BrahMos supersonic cruise missile" in a "major operational lapse" that could have provoked Pakistani retaliation.[58] "Defense sources" also assured that the Indian Court of Inquiry would be "swift" and "fix responsibility for acts of omission and commission" while also reviewing "existing operational and command and control procedures" to prevent any such recurrence.

> The defence establishment remained tight-lipped...but sources said the conventional (non-nuclear) land-attack BrahMos was accidentally launched during a missile unit's training and maintenance exercise near an IAF base in northern India....there was, incidentally, an audit by IAF's Directorate of Air Staff Inspection (DASI) under way at that time. There are multiple mechanical and electronic locks built into BrahMos missiles which fly at almost three times the speed of sound at Mach 2.8....sources said the in-built locks in the BrahMos missile in question were probably accidentally bypassed by operators in the training exercise or there was a technical malfunction. The target coordinates and data, with the requisite waypoints, fed into the missile during such an exercise would also be realistic. But it was certainly not meant to be launched...Once the air-breathing missile took-off, there was no way to abort it. Conventional missiles like the BrahMos do not have self-destruct mechanisms like the ones on strategic or nuclear missiles like the Agni and Prithvi series of ballistic missiles.[59]

Dissatisfied Pakistan Calls for a Joint Probe—Suspects Deliberate Launch

The lack of specifics in the brief statement by the Indian Defence Ministry on 11 March did not satisfy Pakistan with then National Security Adviser, Moeed Yusuf, criticising India for the delay in its clarification in an 11 March X (previously Twitter) thread. Yusuf reckoned that the incident "raises serious questions about India's ability to handle such sensitive technology" as the missile threatened civilians on the ground as well as international and domestic civilian aviation . He also castigated India as "highly irresponsible" for not having notified Pakistan immediately that an "inadvertent launch of a cruise missile had taken place." The Pakistanis had hitherto preferred the terms "supersonic flying object," "Indian projectile," "supersonic projectile" and "projectile" and had conjectured that it was "most probably a missile." Yusuf's statement was the first time that Pakistan clearly identified the intruding projectile as a cruise missile. The Pakistani strategy of using the incident to question India's image as a responsible nuclear weapons state (and as a responsible state, more broadly) also became clear in Yusuf's criticisms.

The statement also highlighted Pakistan's deep suspicions of the Indian government's motives, leading to doubts regarding whether the launch was an accident or a deliberate Indian ploy to test Pakistan's air defenses.

> In a nuclear environment, such callousness & ineptitude raises questions about the safety & security of Indian weapon systems....Remember, this is a state apparatus being run by a fascist ideology that has already proven its recklessness by trying to bomb Pakistan in 2019....It is hard to believe anything this Indian government says. Therefore, the real circumstances surrounding this incident must also be investigated to ascertain if this was an inadvertent launch or something more intentional.[60]

Reinforcing Yusuf's suspicions of the missile launch as deliberate, Pakistani Foreign Minister Shah Mahmood Qureshi also alleged that the launch was an aggressive Indian provocation and not a technical failure as claimed by India and should be seen in the context of the Indian air strike in Balakot in 2019. Enjoining the international community and aviation bodies to take note, Qureshi informed that Pakistan's subsequent steps would involve briefing the envoys of the P-5 countries about the incident even as it waited for a more detailed Indian explanation.[61]

Meanwhile, anonymous "analysts and military officials" in India continued to speculate in the media as to why the missile's self-destruct option was not activated once its accidental launch became known. Their ruminations contradict previous leaks to the media by anonymous "defense sources" (such as the 13 March *Times of India* article covered earlier in the chapter) that the BrahMos did not have a self-destruct mechanism once it was launched.

> Other analysts and military officials speculated as to why the "self-destruct" or re-programming option in mid-flight, available on the BrahMos was not activated, further reinforcing the belief that the missile had gone "totally haywire and out its IAF operators" control. They conjectured whether the "delinquent" missile could alternately have been a Prithvi SSM variant fitted with inertial navigation and ground-based mid-course correction systems, which were believed to be stationed in the area as part of the Strategic Forces Command (SFC) assets, but under the IAFs operational control. But senior military officials and missile experts said India never conducted any Prithvi firings "inland", executing them only from the coastal Integrated Test Range at Balasore in Odisha, over the sea.[62]

The Pakistani Foreign Office released a statement on 12 March in which it expressed dissatisfaction at the "simplistic explanation proffered by the

Indian authorities." Highlighting the "grave nature" of the incident and the questions it raised regarding the integrity of "security protocols and technical safeguards against accidental or unauthorized launch of missiles in a nuclearized environment," the statement made seven demands that would come to constitute Pakistan's response to subsequent Indian clarifications. These were:

- India must explain the measures and procedures in place to prevent accidental missile launches and the particular circumstances of this incident.
- India needs to clearly explain the type and specifications of the missile that fell in Pakistani territory.
- India also needs to explain the flight path/trajectory of the accidentally launched missile and how it ultimately turned and entered into Pakistan.
- Was the missile equipped with self-destruct mechanism? Why did it fail to actualize?
- Are Indian missiles kept primed for launch even under routine maintenance?
- Why did India fail to immediately inform Pakistan about accidental launch of the missile and waited to acknowledge till after Pakistan announced the incident and sought clarification?
- Given the profound level of incompetence, India needs to explain if the missile was indeed handled by its armed forces or some rogue elements?[63]

The Foreign Office statement also noted the "short distances and response times" in South Asia and the potential for misinterpretation and miscommunication to lead to escalation with "grave consequences."[64] It ended by calling for a joint probe into the incident instead of an Indian Court of Inquiry given the missile's landing in Pakistani territory.[65] The Foreign Office also summoned India's Charge d'Affaires (Cd'A) on the same day and reiterated its call for satisfactory answers to its seven questions and for a joint probe.[66] While addressing a political rally in Punjab's Hafizabad district, then Pakistan Prime Minister Imran Khan weighed in for the first time on India's accidental missile launch and quipped that Pakistan could have responded to India's missile but chose to exercise restraint.[67]

Meanwhile, US State Department spokesperson Ned Price supported the Indian position of the missile launch being accidental in response to a question by a Pakistani journalist during a press briefing.[68] Pakistan's Foreign Minister Shah Mehmood Qureshi continued to reiterate the demand for a joint probe into the incident and the need for the international community to play "its due role" in promoting "strategic stability" in his telephonic conversation with the German Foreign Minister and the United Nations Secretary General on 14 March.[69]

India's Second Statement—from Technical Malfunction to Human Error

It was on 15 March that India made public its first detailed response to the missile accident. Defense Minister Rajnath Singh came up with a more detailed, identical statement in both houses of the Indian Parliament. He repeated the Indian explanation of the missile being accidentally released on 11 March 2022 at approximately 19:00 "during routine maintenance and inspection of a missile unit." Terming the incident as "regrettable," he expressed relief that the incident "did not cause any kind of loss." Singh assured that the government had taken the incident "very seriously" and had ordered a "high-level Court of Enquiry." Singh went beyond India's terse 11 March statement by assuring that the "standard operating procedures for operation, maintenance (and) instruction are also being reviewed" and any gaps therein would be immediately redressed.

Without revealing the specifics of the missile that had crashed in Pakistan, Singh gave a generic set of reassurances that "our missile system is extremely safe and reliable," India's safety protocols "are of a high standard" and that India's armed forces were "well-trained and disciplined and...well-experienced in the handling of these types of systems."[70] Singh's statement also did not reveal the missile's launch site or the branch of the military (army, navy, airforce) that was involved in the accidental release.[71] Senior service officers, speaking on the condition of anonymity to the press were quick to notice Singh's omission of "technical malfunction" mentioned in the Defence Ministry's 11 March press release and speculated instead that his focus on military standard operating procedures, pointed to human error rather than technical error.[72]

> The BrahMos flew into Pakistan, after the requisite waypoints, since the target coordinates fed into the missile were also realistic in the full-scale simulation. "It was an operational blunder by some overenthusiastic operators during the inspection. BrahMos is a fool-proof missile with several in-built mechanical and electronic locks to prevent accidental launches... they were obviously also by-passed", a source said.[73]

The Aborted Pakistani Retaliatory Strike, Call for Joint Probe

Singh's statement to Parliament was scant on details and its generic reassurances elicited concerted critique from a dissatisfied Pakistan. Foreign Minister Shah Mahmood Qureshi rejected Singh's statement as insufficient and reiterated the Pakistani demand for a joint probe. Qureshi highlighted that the Indian missile was capable of carrying a nuclear warhead and could have resulted in a Pakistani retaliation.

> In his comments, Qureshi lamented that the international community had failed to realize the severity of the Indian missile launch as it was capable of carrying the nuclear warhead and could have resulted in a nuclear conflagration. "One sided investigation is not enough; its impact could be far beyond the region, because the missile was capable of carrying a warhead. It had this potential (of carrying a warhead) and to say that it was an accident, it is not enough," he said. "There could be an accidental reaction to that accidental launch if our air force had not picked it up well inside India. Do people realize the implication and consequences of that? That is very serious," he said.[74]

Qureshi also informed that he had written to the UN Secretary General and the President of the Security Council about the incident.[75] Pakistani National Security Adviser Moeed Yusuf also contended that India's "simplistic explanation without evidence" that the accident occurred during "routine maintenance" should be "unacceptable to the world." Yusuf reminded that the missile in question was a "highly sophisticated supersonic missile" which could have resulted in "escalation between two nuclear-armed countries." He further claimed that Pakistan had acted responsibly and avoided escalation in a manner akin to its voluntary return of a pilot whose aircraft it shot down during India's airstrikes following the Pulwama attack of 2019. Yusuf then characterized India as an "irresponsible and untrustworthy state when it comes to handling of such sensitive weapon systems" and concluded that "only a transparent joint probe can address the many unanswered questions about this so-called mistake."[76] A 15 March article by *Bloomberg*, quoted in the Indian media, cited unnamed sources warning that Pakistan had initially contemplated a retaliatory strike with a similar missile in the aftermath of the accident but held back.

> A day after Defence Minister Rajnath Singh explained to parliament how a missile was "inadvertently launched" on March 9 and landed in Pakistan, a *Bloomberg* report claims that Pakistan had prepared to launch a similar missile in a retaliatory strike. Pakistan held back because an initial assessment indicated something was amiss, *Bloomberg* quoted unnamed sources as saying.[77]

Pakistan's army top-brass also weighed in on the incident during the 248th Corps Commanders Conference held at GHQ, Rawalpindi on 15 March, presided over by Chief of Army Staff, General Qamar Javed Bajwa. Warning that the "recent incident of missile firing, claimed as accidental by India could have resulted in a major disaster" the conference called on "relevant international forums" to "subject the safety and security protocols of Indian strategic assets to deliberate oversight."[78]

On the Indian side, an article in *The Tribune* claiming access to Indian official sources informed that a BrahMos missile had been launched on 9 March at 19:00 from a base near Ambala (and not Sirsa as claimed by Pakistan) and that a "high-level inquiry team will interact with a clutch of Indian Air Force (IAF) officers posted in and around Ambala and examine missile storage and inspection protocol." [79]

Claiming that the inquiry team would include a representative from the Indian Air Force, the article also informed that "IAF officers in the chain of command will be called to find out why the missile was fired and how it landed in Pakistan. Each data, log and procedure followed by the IAF officials on the day is expected to be scrutinized."[80] The article also claimed that the accident occurred due to "human and technical errors during a routine exercise to check how systems would function in case of war" and that the Ministry of Defence had already ordered a "review of standard storage procedures, maintenance and inspecting manuals."[81]

Even as more details about the missile accident continued to emerge in the Indian media, a Pakistani spokesperson continued to reiterate Pakistan's demand for a joint probe and answers to the seven questions it posed to India earlier in press briefing on 18 March.[82] Foreign Minister Qureshi also continued to internationalize the issue by reiterating his demand for a joint probe in bilateral meetings with the Foreign Minister of Egypt on 20 March,[83] the Chinese Foreign Minister on 21 March[84] and the Visiting US Under Secretary of State, also on 21 March.[85]

Pakistan also succeeded in securing a favorable resolution at the 48th session of the 57-member Organization of Islamic Countries (OIC) on 23 March which endorsed its call for a joint probe, expressed serious concerns over "multiple violations of international law" caused by the incident and called on the UN Secretary General, UN Security Council and International Civil Aviation Authority (ICAO) to take up the matter with India.[86] The favorable OIC resolution was perceived by Pakistani National Security Advisor Moeed Yusuf as increasing "global calls for accountability" and a "transparent joint investigation."[87]

Meanwhile, a 17 March article in *The Wire* by journalist Rahul Bedi channeled the concerns of Indian military planners regarding the technical details of the errant BrahMos missile being passed onto China by Pakistan.

> Indian military planners fear that a forensic examination of the BrahMos missile remains in Mian Channu by Pakistani technicians and defence experts, could to some extent undermine their efforts at fielding a daunting and unique cruise missile which till now was a closely-held secret.

> Moreover, considering Islamabad's military, political and financial ties with Beijing, it's a foregone conclusion that China too will be provided the data Pakistan can glean and deduce from the rogue BrahMos's remains. In short, the prospective 'human error' that might perhaps have led to the unplanned missile launch could, ultimately, prove expensive.[88]

The Court of Inquiry Determines Human Error as Primary Cause

About a week after Rajnath Singh's statement, reports in the Indian press began to provide snippets of information about the ongoing Indian Court of Inquiry. A 24 March article in *The Tribune* reported that the inquiry had "examined the role of at least four Indian Air Force (IAF) officials—a Group Captain (Commanding Officer), his seniors and also his juniors" and had determined that human error was the primary cause of the incident. The probe's focus on how the launch codes were entered into the system that led to the firing of the missile also indicates that the human error angle was being examined as the primary cause by this point in time.

> The initial probe has indicated a "human error", and not a technical fault, led to the missile's launch. Though alleged lapses in handling the missile are being probed, the CoI is looking at multiple angles. All missile storage and inspecting protocols have been examined. IAF officials in the chain of command have been questioned to find out why and how the missile was fired. Each data, log and procedure followed by the officials on March 9 has been scrutinised by the CoI. The key question is how the launch codes were fed into the system that enabled the missile to fire. The protocol for firing and storage of BrahMos, a supersonic cruise missile, is such that a series of steps are needed to be completed before actual firing.[89]

A revealing article by veteran journalist Pradip Sagar published a day earlier in *The Week* magazine reported on the IAF leadership heading the probe and the four-hour, usual preparatory process to launch the BrahMos missile. Sagar also inferred the need for regular psychological tests for missile unit personnel from the statements of anonymous defense officials. Especially insightful was his claim that the Indian Defence Ministry's initial statement on 11 March of a "technical malfunction" leading to the missile accident was inadvertent and BrahMos missile developers were upset by the phrasing.

> The inquiry into the misfire incident of an Indian missile into Pakistan has pointed towards "human error". An air vice-marshal, a two-star rank officer of the air force is heading the probe, which is expected to be completed in a month and is mandated to fix responsibility for acts of omission and commission. It appears several officials handling the missile

> system are likely to be indicted for the lapse. "It's a case of immature operation or handling (mishandling) of the missile system, as there is no question of missile missing its target," a key defence official said. Another officer said the process takes at least four hours to launch the missile with multiple electronic and mechanical locks. "Deviating from its flying path (trajectory) is next to impossible in case of BrahMos," said an officer, indicating towards 'human error' for the misfire incident. The incident also drew attention to the need to carry out regular psychological tests and background checks for all the officials handling such strategic arsenal.... Earlier, India's ministry of defence had issued a statement, in which, it had inadvertently mentioned that "in the course of a routine maintenance, 'a technical malfunction' led to the accidental firing of a missile". The word "technical malfunction" has upset the developers of the BrahMos missile.[90]

The aforementioned reportage raises some interesting questions. Was the Defence Ministry's 11 March statement identifying technical error (i.e. technical malfunction) as the main cause of the missile accident indeed inadvertent? And was Rajnath Singh's subsequent statement to Parliament on 15 March in which the technical malfunction phrase was absent (redirecting the focus of attention to standard operating procedures and the human error angle) indeed the result of a pushback from the BrahMos missile developers trying to defend the technical capabilities of their missile (and by extension their own technical prowess)? The documentary record does not contain additional evidence to further develop this line of inquiry.

Another article on 23rd March also conveyed that an Air Vice Marshal (confirmed in a later press article as Assistant Chief of Air Staff Operations (Offensive) Air Vice Marshal RK Sinha[91]) had ascertained that human error was to blame for the accidental release with a Group Captain being held primarily responsible.

> Government sources told *ANI* that the detailed probe by the Air Force officer (equivalent to Major General in the Army) is still going on but prima facie, a Group Captain-rank officer is being seen as blameworthy. The officer was in charge of the mobile command post of the missile system when the accidental firing happened during a Command Air Staff Inspection (CASI) at its home base. However, the sources said the probe would be finalised in a time-bound manner and final details would be known only after that. The sources said the AVM (Air Vice Marshal) probing the incident is highly qualified and is in-charge of operations in the Air Force headquarters.[92]

The *Times of India* detailed in an article on 24 March that the "tactical missile was launched by mistake during a simulation exercise being audited by a Command Air Staff Inspection" with a Group Captain in charge of the mobile command post of the BrahMos unit.[93] The article also conveyed that "there is no way to abort tactical missiles like the BrahMos once they are launched because they do not have self-destruct mechanisms."[94] Another article on the same day in *BW Businessworld* informed that the Indian Court of Inquiry was being conducted by a "two-star officer"[95] and that the investigation was proceeding at a quick pace as "there's pressure on the IAF to (take) swift demonstrable punitive action and reinforce safety measures in revised SOPs."[96]

Even as the Court of Inquiry's investigations were underway, the IAF began efforts to bolster its image that had been damaged by the accidental launch of 9 March. In its first test since the accidental launch, the IAF conducted a successful test of a "new version of a land attack BrahMos missile" at the Andaman and Nicobar Command (ANC) with an extended strike range of 350–400 km which hit the target "with pinpoint accuracy," in the presence of both the IAF Air Vice Marshall and the ANC Chief.[97]

The Indian government also sought to reassure the Philippines, which had signed a $375 million deal to acquire BrahMos missiles that there was no technical defect in the missile

> A report in *Indian Express* said Philippines Defence Secretary Delfin Lorenzana had called in Indian ambassador Shambhu S Kumaran who assured the senior minister that there was no technical issue with the missile system that is manufactured by BrahMos Aerospace in India. Kumaran is also reported to have told Lorenzana that an enquiry into the incident is underway and probe details will be shared with relevant authorities. While there has been speculation to this effect, there is no official confirmation that the missile accidentally fired was a BrahMos..... In the first-ever export order for BrahMos missiles, Philippines had signed $374 million contract with the BrahMos Aerospace Pvt Ltd (BAPL) in January this year to supply an undisclosed number of missiles.[98]

The Court of Inquiry Completes Its Probe; Three Officers Dismissed

After a brief lull, an 11 April article conveyed that the IAF had completed its inquiry and "held more than one official from the missile squadron responsible for the accident," constituting "a clear violation of standard operating procedures."[99] The article further updated that the "Defense Ministry has also been briefed about the IAF inquiry committee report" and "action is expected in the next few weeks."[100] Another article published a day earlier, channeled the probe's key conclusion terming the accidental launch as "totally avoidable," identifying "more than one official from the missile squadron" as responsible and promising "swift and severe punishment."[101]

> The Commanding Officer (CO) of an Indian Air Force (IAF) BrahMos missile command post which accidentally fired a missile into Pakistan on March 9 has reportedly been found blameworthy for grave operational lapses by a high-level court of inquiry. Reports indicate that "more than one official" has been indicted in the inquiry report. Reports, which have not been officially corroborated by the IAF, suggest that the court of inquiry has ruled out technical malfunction, as was initially claimed. It appears that the inbuilt mechanical and electronic locks of the 290 km range supersonic missile system had been overridden as a consequence of the violation of SOPs.[102]

The Indian Court of Inquiry had been completed in little over a month following the 9 March missile accident, a surprising achievement in a country often known for incremental bureaucratic decision-making. The reason for the swift conclusion of the inquiry was the strong intent within the IAF top brass and the government embarrassed by the missile accident to hold those accountable for the loss of face and for damage to India's carefully cultivated reputation as a responsible state.

> The government and Indian Air Force top brass are of the view that the punishment should be quick and not allowed to prolong which has happened in many cases earlier, sources said. The inquiry also brought out the role of an official in controlling the situation after the missile got launched accidentally. The IAF official probing the incident was asked to complete the inquiry into the incident in the earliest possible timeframe as accountability of officials had to be fixed for the mishap which could have caused much more damage. Indian Air Force has taken the incident very seriously and taking every step internally to ensure that such incidents don't occur in future and various standard operating procedures are being reviewed to see if there is any change required to make things smoother, the sources said.[103]

The Defence Ministry's final decision on the findings of the Court of Inquiry became known on 23 August with an article reporting that the IAF had terminated the services of a Group Captain and two Wing Commanders with immediate effect.

> The government on Tuesday terminated the services of three Indian Air Force (IAF) officers after finding them blameworthy for the accidental firing of a BrahMos missile from a base near Ambala on March 9.....The Air Force said the three officers had "primarily been held responsible for the incident". The initial probe indicated a "human error", and not a technical fault, had led to the fire. The CoI found the Commanding

> Officer (Group Captain) and others "blameworthy of serious operational lapse". The IAF decided to terminate the services of the three officers as the matter was seen a serious lapse capable of triggering a war between India and Pakistan.[104]

A 24 August article explained that, following the conclusion of its inquiry into the incident, the "IAF in April changed all SOPs for handling of missiles."[105]

BrahMos Aerospace Private Limited, the manufacturer of the BrahMos missile had maintained a conspicuous silence ever since the 9 March accident. In his first public statement on the accidental fire, BrahMos Aerospace Director Atul Rane expressed satisfaction over the COI's findings that absolved the company of any blame for the accident by ruling out the possibility of a technical malfunction.

> A court of inquiry was set up and it was concluded that the SOPs were not followed by the three IAF officers. You may term it as riding a bicycle without hands," Rane told Raj Chengappa, Group Editorial Director (Publishing), India Today Group at the BT's India@100 summit in New Delhi on Friday. He added that the company has proper SOPs in place, and there has been no technical default at the company's end.[106]

Pakistan Persists with Joint Probe

The relatively quick conclusion of the Indian probe and its dismissal of three IAF officers for deviation from the SOPs had fallen well short of the Pakistani demand for a joint probe. The Pakistani Ministry of Foreign Affairs released a statement on 24 August rejecting the COI's findings and reiterating its call for a joint probe.

> As expected, the measures taken by India in the aftermath of the incident and the subsequent findings and punishments handed by the so-called internal Court of Inquiry are totally unsatisfactory, deficient and inadequate. India has not only failed to respond to Pakistan's demand for a joint inquiry but has also evaded the questions raised by Pakistan regarding the command and control system in place in India, the safety and security protocols and the reason for India's delayed admission of the missile launch. Systemic loopholes and technical lapses of serious nature in handling of strategic weapons cannot be covered up beneath the veneer of individual human error.[107]

The Pakistani reaction to the Indian probe indicated its suspicion that India was trying to cover up technical defects in the missile system, organizational deficiencies in its command and control system and procedural shortfalls in its standard operating procedures by pinning the blame for the accident on the three IAF officers.

The identity of one of three dismissed IAF officers became clear on 1 March 2023, seven months after the dismissal when Wing Commander Abhinav Sharma filed a case against India's Defence Ministry and the Chief of the Indian Air Force (IAF) in the Delhi High Court claiming wrongful dismissal.[108] The specific nature of the allegations made by the Court of Enquiry against Sharma and the government's response was also revealed in media coverage of the case, which is focused on in the Conclusion of this book.

Analysis—Maximalist Pakistan and Minimalist India

A constructivist analysis of the response of the Indian and Pakistani governments to the missile accident reveals interesting insights. It highlights the initial period of confusion in Pakistan regarding the exact nature of the intruding flying object, the evolution in the Pakistani government's response from the cautious tone struck in the press conference of 9 March to its strongly worded criticisms of the Indian government's motives, standard operating procedures (SOPs) and demands for a joint probe. These criticisms tended to adopt a maximalist interpretation of the potential risks from the event (the missile being armed with a warhead, the missile landing in a populated area causing casualties, the missile slamming into domestic flights that were airborne at the time along its flight-path, the missile being supposedly capable of carrying a nuclear warhead and the prospect of a deliberate launch by a purportedly fascist Indian government etc). The evolution in the Pakistani tone from initially cautious to subsequently assertive points to the subsuming of the missile misfire incident into the broader geopolitical and security competition dynamic between India and Pakistan with Pakistan focusing on denting India's image as a responsible state which had allowed it to de-hyphenate itself from Pakistan, secure US recognition (albeit indirect) for its nuclear weapons under the US-India nuclear deal, enter various export control regimes and secure nuclear and dual-use technologies from the West.

A constructivist perspective is also useful in tracing the evolution in the Indian government's statements from its initial identification of technical malfunction as the main reason for the accident in its 11 March statement to its shift towards the human error rationale in its 15 March statement and the eventual solidification of the human error rationale in the preliminary and final results of the Court of Inquiry. In stark contrast to the Pakistani government's statement, India's official responses tended to adopt a minimalist interpretation of the accident (emphasis on the missile being misfired accidentally due to a technical malfunction, relief at the accidental launch not causing any Pakistani casualties, confidence in the ability of the armed forces to handle sophisticated weapons systems, confidence in the Court of Inquiry for conducting a transparent investigation and fixing accountability in a timebound manner etc). On a broader level, Indian actions seemed geared towards demonstrating to key domestic constituencies (national security community consisting of strategic experts, veteran journalists, opposition party leaders, current and former military officials etc) and to the international community India's capacity to quickly affix responsibility for mishandling of sophisticated weapons systems in a timebound manner as also its ability to overhaul standard operating procedures to prevent a recurrence of the missile misfire.

A fleeting glimpse of the internal pulls and pressures within the Indian system emerges from a granular examination of the discourse on the Indian side as witnessed in the claimed dissatisfaction among the designers of the BrahMos following the technical error rationale advanced by the Indian government and the satisfaction expressed by the Director of BrahMos Aerospace Private Limited regarding the human error rationale later identified as the primary cause by the Court of Inquiry (thus eliminating the technical error rationale as a probable cause).

Notes

1 Snehesh Alex Philip, "Missile Accidentally Fired 'during Inspection' at Secret IAF Base, Pakistan 'Didn't Track,'" *ThePrint*, March 12, 2022, accessed December 3, 2022, https://theprint.in/defence/missile-accidentally-fired-during-inspection-at-secret-iaf-base-pakistan-didnt-track/870814/, Daryl Kimball, "India Accidentally Fires Missile Into Pakistan," *Arms Control Association,* April 2022, accessed December 3, 2022, https://www.armscontrol.org/act/2022-04/news/india-accidentally-fires-missile-into-pakistan

2 Amitabh Sinha, "Misfire of BrahMos Missile Not Cause for Any Specific Concern: IAEA," *The Indian Express*, November 14, 2022, accessed December 3, 2022, https://indianexpress.com/article/world/misfire-of-brahmos-missile-not-cause-for-any-specific-concern-iaea-8266709/

3 Daryl Kimball, "India Accidentally Fires Missile Into Pakistan," *Arms Control Association,* April 2022, accessed December 3, 2022, https://www.armscontrol.org/act/2022-04/news/india-accidentally-fires-missile-into-pakistan

4 Matt Korda, "Flying Under The Radar: A Missile Accident in South Asia," *Federation of American Scientists,* April 4, 2022, accessed December 6, 2022, https://fas.tghp.co.uk/publication/flying-under-the-radar-a-missile-accident-in-south-asia/

5 Vishnu Som, "Indian Missile Accident Nearly Led to Pak Retaliation: Report," *NDTV.com,* March 16, 2022, accessed December 6, 2022, https://www.ndtv.com/india-news/indian-missile-accident-nearly-led-to-pak-retaliation-report-2825860

6 Ranjan Sudhi Sen and Mangi Faseeh, "Errant Indian Missile Nearly Led to Pakistan Retaliatory Strike," *Bloomberg.com,* March 15, 2022, accessed December 7, 2022, https://www.bloomberg.com/news/articles/2022-03-15/errant-indian-missile-nearly-led-to-pakistan-retaliatory-strike

Sajjad Hussain, "Pakistan's New Security Policy Seeks 'Peace' with India: Report," *The Economic Times*, January 12, 2022, accessed December 7, 2022, https://economictimes.indiatimes.com/news/defence/pakistans-new-security-policy-seeks-peace-with-india-report/articleshow/88855569.cms?utm_source=contentofinterest&utm_medium=text&utm_campaign=cppst

7 Elizabeth Roche, "Pakistan Postpones SAARC Summit after India-Led Boycott over Uri Attack," *Mint,* September 30, 2016, accessed December 7, 2022, https://www.livemint.com/Politics/waHWF3HUajmf66JdNwnCqJ/Pakistan-cancels-Saarc-Summit-after-Indialed-boycott-over-U.html

8 "Surgical Strike Day: Here's How the 2016 Operation Was Carried Out," *Hindustan Times*, September 29, 2021, accessed December 8, 2022, https://www.hindustantimes.com/india-news/surgical-strike-day-here-s-how-the-2016-operation-was-carried-out-101632882272993.html

9 Riya Teotia, "Explained: Times When Pulwama Terror Attack Became a Flashpoint in Indian Politics," *WION*. April 19, 2023, accessed October 19, 2023, https://www.wionews.com/india-news/explained-times-when-pulwama-terror-attack-became-a-flashpoint-in-indian-politics-583788

Livemint, "4 Years of Pulwama Attack: From Balakot Airstrike to Fears of Nuclear War," *Mint,* February 14, 2023, accessed October 19, 2023, https://www.livemint.com/news/india/4-years-of-pulwama-attack-from-balakot-air-strike-to-fears-of-nuclear-war-11676349092294.html

10 Bureau, "Pulwama Terror Attack Anniversary: Timeline of How Terrorist Attacked CRPF Convoy and Indias Response With Balakot Airstrike," *Zee News,* February 14, 2023, accessed October 19, 2023, https://zeenews.india.com/india/pulwama-terror-attack-anniversary-timeline-of-how-terrorist-attacked-crpf-convoy-and-indias-response-with-balakot-airstrike-2573084.html

Riya Teotia, "Balakot Airstrikes: How Indian Air Force Carried out a Successful Retaliation Four Years Back," *Wion*, February 25, 2023, accessed October 20, 2023, https://www.wionews.com/india-news/balakot-airstrikes-how-india-air-force-carried-out-a-successful-retaliation-four-years-back-565895

11 Amit Chaturvedi, "2 Years of Abrogation of Article 370: 5 Big Changes in Jammu and Kashmir," *Hindustan Times,* August 5, 2021, accessed December 10, 2022, https://www.hindustantimes.com/india-news/2-years-of-abrogation-of-article-370-5-big-changes-in-jammu-and-kashmir-101628130402726.html

12 Mehmal Sarfraz, "Kashmir Issue: Pakistan Expels Indian Envoy, Suspends Bilateral Trade," *The Hindu,* August 7, 2019, accessed December 11, 2022, https://www.thehindu.com/news/international/kashmir-issue-pakistan-expels-indian-envoy-suspends-bilateral-trade/article61586712.ece

13 PIB Delhi, "Joint Statement," *Ministry of Defence*, India. February 25, 2021, accessed December 11, 2022, https://pib.gov.in/pib.gov.in/Pressreleaseshare.aspx?PRID=1700682

14 Surya Valliappan Krishna, "Bordering on Peace: Evaluating the Impact of the India-Pakistan Ceasefire," *Carnegie India,* February 24, 2022, accessed December 12, 2022, https://carnegieindia.org/2022/02/24/bordering-on-peace-evaluating-impact-of-india-pakistan-ceasefire-pub-86513 Naseer Ganai, "India-Pakistan LOC Truce: One Year on, Peace Prevails," *Outlook,* February 26, 2022, accessed December 12, 2022, https://www.outlookindia.com/national/indo-pak-loc-truce-one-year-on-peace-prevails-news-184182

15 Outlook Web Desk, "Pakistan Willing to Make Peace with India under New National Security Policy: Report," *Outlook*, January 12, 2022, accessed December 12, 2022, https://www.outlookindia.com/website/story/india-news

-pakistan-willing-to-make-peace-with-india-under-new-national-security-policy-report/409404
Sameer Patil, "Pakistan's New Security Policy and Bid for Peace with India," *ORF*, January 21, 2022, accessed December 13, 2022, https://www.orfonline.org/research/pakistans-new-security-policy-and-bid-for-peace-with-india/

16 New Delhi, "Exchange of List of Prisoners between India and Pakistan," *Ministry of External Affairs*, January 1, 2022, accessed December 13, 2022, https://www.mea.gov.in/press-releases.htm?dtl/34735/Exchange_of_list_of_prisoners_between_India_and_Pakistan

17 PTI, "Trade with India Need of the Hour: Adviser to Pakistan PM," *Business Today*, February 21, 2022, accessed December 13, 2022, https://www.businesstoday.in/latest/economy/story/trade-with-india-need-of-the-hour-adviser-to-pakistan-pm-323337-2022-02-21

18 Sachin Parashar, "India, Pakistan, China among 35 Nations to Abstain from Voting on Anti-Russia Resolution," *The Times of India*, March 3, 2022, accessed December 12, 2022, https://timesofindia.indiatimes.com/world/europe/india-pakistan-china-among-35-nations-to-abstain-from-voting-on-anti-russia-resolution/articleshow/89955078.cms
Neeta Sharma, "India And Pak Willing To Walk, If Not Work, Together: Government Source," *NDTV.com*, March 4, 2022, accessed December 13, 2022, https://www.ndtv.com/india-news/india-and-pak-willing-to-walk-if-not-work-together-government-source-2804226

19 New Delhi, "117th Meeting of the India-Pakistan Permanent Indus Commission," *Ministry of External Affairs*, Government of India. March 3, 2022, accessed December 13, 2022, https://www.mea.gov.in/press-releases.htm?dtl/34925/117th+meeting+of+the+IndiaPakistan+Permanent+Indus+Commission

20 News Desk, "Fighter Jet Crashes in Mian Channu," *The Nation*, March 9, 2022, accessed December 13, 2022, https://nation.com.pk/2022/03/09/fighter-jet-crashes-in-mian-channu/

21 Web Desk, "Aircraft Crashes in Mian Channu, No Causalities Reported," *ARY NEWS*, March 9, 2022, accessed December 15, 2022, https://arynews.tv/aircraft-crashes-mian-channu/

22 Staff Reporter, "Trainer Crashes in Mian Channu," *DAWN.COM*, March 10, 2022, accessed December 15, 2022, https://www.dawn.com/news/1679164

23 Staff Reporter, "Trainer Crashes in Mian Channu," *DAWN.COM*, March 10, 2022, accessed December 15, 2022, https://www.dawn.com/news/1679164

24 DG ISPR Press Conference. *Official DGISPR*, March 10, 2022, accessed December 15, 2022,
DG ISPR on Twitter: "DG ISPR Press Conference - 10 March 2022 https://t.co/W5HTbFll3V" / Twitter
DG ISPR Lt Gen Babar Iftikhar Important Press Conference. *Public News*, March 10, 2022, accessed December 15, 2022, https://www.youtube.com/watch?v=ddW7rb-xGTo

25 DG ISPR Press Conference. *Official DGISPR*, March 10, 2022, accessed December 15, 2022, DG ISPR on Twitter: DG ISPR Press Conference - 10 March 2022 https://t.co/W5HTbFll3V" / Twitter DG ISPR Lt Gen Babar Iftikhar Important Press Conference. *Public News*, March 10, 2022, accessed December 15, 2022, https://www.youtube.com/watch?v=ddW7rb-xGTo

26 DG ISPR Press Conference. *Official DGISPR*, March 10, 2022, accessed December 15, 2022, DG ISPR on Twitter: DG ISPR Press Conference - 10 March 2022 https://t.co/W5HTbFll3V" / Twitter DG ISPR Lt Gen Babar Iftikhar Important Press Conference. *Public News*, March 10, 2022, accessed December 15, 2022, https://www.youtube.com/watch?v=ddW7rb-xGTo

27 DG ISPR Press Conference. *Official DGISPR,* March 10, 2022, accessed December 15, 2022, DG ISPR on Twitter: DG ISPR Press Conference - 10 March 2022 https://t.co/W5HTbFll3V" / Twitter DG ISPR Lt Gen Babar Iftikhar Important Press Conference. *Public News,* March 10, 2022, accessed December 15, 2022, https://www.youtube.com/watch?v=ddW7rb-xGTo

28 MFA, "Pakistan Registers Strong Protest over Unprovoked Violation of Its Airspace by a 'super-Sonic Flying Object' of Indian Origin," *Ministry of Foreign Affairs*, March 11, 2022, accessed December 15, 2022, https://mofa.gov.pk/pakistan-registers-strong-protest-over-unprovoked-violation-of-its-airspace-by-a-super-sonic-flying-object-of-indian-origin-2/

29 MFA, "Pakistan Registers Strong Protest over Unprovoked Violation of Its Airspace by a 'super-Sonic Flying Object' of Indian Origin," *Ministry of Foreign Affairs*, March 11, 2022, accessed December 15, 2022, https://mofa.gov.pk/pakistan-registers-strong-protest-over-unprovoked-violation-of-its-airspace-by-a-super-sonic-flying-object-of-indian-origin-2/

30 MFA, "Pakistan Registers Strong Protest over Unprovoked Violation of Its Airspace by a 'super-Sonic Flying Object' of Indian Origin," *Ministry of Foreign Affairs*, March 11, 2022, accessed December 15, 2022, https://mofa.gov.pk/pakistan-registers-strong-protest-over-unprovoked-violation-of-its-airspace-by-a-super-sonic-flying-object-of-indian-origin-2/

31 Alex Snehesh Philip and ThePrint, "Was It a Missile? Was It Brahmos? India Probing Pakistani Claim of Airspace Intrusion," *ThePrint*, March 11, 2022, accessed December 15, 2022, https://theprint.in/defence/was-it-a-missile-was-it-brahmos-india-probing-pakistani-claim-of-airspace-intrusion/868536/?amp

32 PIB Delhi, "Statement on Accidental Firing of Missile," *Ministry of Defence.* Government of India, March 11, 2022, accessed December 15 2022. https://pib.gov.in/pib.gov.in/Pressreleaseshare.aspx?PRID=1805148

33 PIB Delhi, "Statement on Accidental Firing of Missile," *Ministry of Defence.* Government of India, March 11, 2022, accessed December 15 2022. https://pib.gov.in/pib.gov.in/Pressreleaseshare.aspx?PRID=1805148

34 PIB Delhi, "Statement on Accidental Firing of Missile," *Ministry of Defence.* Government of India, March 11, 2022, accessed December 15 2022. https://pib.gov.in/pib.gov.in/Pressreleaseshare.aspx?PRID=1805148

35 Daryl Kimball, "India Accidentally Fires Missile Into Pakistan," *Arms Control Today.* Arms Control Association, April 2022, accessed December 15, 2022, https://www.armscontrol.org/act/2022-04/news/india-accidentally-fires-missile-into-pakistan

36 Matt Korda, "Flying Under The Radar: A Missile Accident in South Asia," *Federation of American Scientists,* April 4 2022, accessed December 16, 2022, https://fas.tghp.co.uk/publication/flying-under-the-radar-a-missile-accident-in-south-asia/

37 Matt Korda, "Flying Under The Radar: A Missile Accident in South Asia," *Federation of American Scientists,* April 4 2022, accessed December 16, 2022, https://fas.tghp.co.uk/publication/flying-under-the-radar-a-missile-accident-in-south-asia/

38 Jain Smriti, "BrahMos: Building World's 'Fastest Supersonic Cruise Missile', Future Variants & Export Orders," *The Times of India,* June 14, 2023, accessed October 20, 2023, https://timesofindia.indiatimes.com/business/india-business/brahmos-building-worlds-fastest-supersonic-cruise-missile-future-variants-export-orders/articleshow/100979055.cms?from=mdr

39 Matt Korda, "Flying Under The Radar: A Missile Accident in South Asia," *Federation of American Scientists,* April 4, 2022, accessed December 16, 2022,

https://fas.tghp.co.uk/publication/flying-under-the-radar-a-missile-accident-in-south-asia/
"Pak Missile Misfire: IAF Probe Blames More Than One Official; 'Severe Punishment' to Be Meted Out," *News18*, April 10, 2022, accessed December 16, 2022, https://www.news18.com/news/india/pak-missile-misfire-iaf-probe-blames-more-than-one-official-severe-punishment-to-be-meted-out-4961396.html
Snehesh Alex Philip, "IAF to Increase Sukhois Armed with BrahMos Supersonic Cruise Missile with over 500km-Range," *ThePrint,* October 5, 2022, accessed December 16, 2022, https://theprint.in/defence/iaf-to-increase-sukhois-armed-with-brahmos-supersonic-cruise-missile-with-over-500km-range/1155562/
FP Explainers, "India Accidentally Fires Missile into Pakistan: What and How Did It Happen?," *Firstpost,* March 12, 2022, accessed December 16, 2022, https://www.firstpost.com/india/india-accidentally-fires-missile-into-pakistan-what-and-how-did-it-happen-10452681.html

40 "Joint Venture," - *BrahMos.Com,* accessed December 16, 2022, https://www.brahmos.com/content.php?id=1
FP Explainers, "India Accidentally Fires Missile into Pakistan: What and How Did It Happen?," *Firstpost,* March 12, 2022, accessed December 16, 2022. https://www.firstpost.com/india/india-accidentally-fires-missile-into-pakistan-what-and-how-did-it-happen-10452681.html

41 FP Explainers, "India Accidentally Fires Missile into Pakistan: What and How Did It Happen?," *Firstpost,* March 12, 2022, accessed December 16, 2022. https://www.firstpost.com/india/india-accidentally-fires-missile-into-pakistan-what-and-how-did-it-happen-10452681.html
Jain Smriti, "BrahMos: Building World's 'Fastest Supersonic Cruise Missile', Future Variants & Export Orders," *The Times of India,* June 14, 2023, accessed October 20, 2023, https://timesofindia.indiatimes.com/business/india-business/brahmos-building-worlds-fastest-supersonic-cruise-missile-future-variants-export-orders/articleshow/100979055.cms?from=mdr

42 Rahul Bedi, "India Needs to Come Clean on 'Accidental Firing of Missile' to Avert Conflict with Pakistan," *The Wire*, March 12, 2022, https://thewire.in/security/india-needs-to-come-clean-on-accidental-firing-of-missile-to-avert-conflict-with-pakistan
FP Explainers, "India Accidentally Fires Missile into Pakistan: What and How Did It Happen?," *Firstpost,* March 12, 2022, accessed December 16, 2022, https://www.firstpost.com/india/india-accidentally-fires-missile-into-pakistan-what-and-how-did-it-happen-10452681.html

43 Joy Mitra, "Nuclear BrahMos: On the Anvil?" *Stimson Center,* September 18, 2020, accessed December 17, 2022, https://www.stimson.org/2020/nuclear-brahmos-on-the-anvil/

44 FP Explainers, "India Accidentally Fires Missile into Pakistan: What and How Did It Happen?," *Firstpost,* March 12, 2022, accessed December 16, 2022, https://www.firstpost.com/india/india-accidentally-fires-missile-into-pakistan-what-and-how-did-it-happen-10452681.html

45 "Pak Missile Misfire: IAF Probe Blames More Than One Official; 'Severe Punishment' to Be Meted Out," *News18,* April 10, 2022, accessed December 16, 2022, https://www.news18.com/news/india/pak-missile-misfire-iaf-probe-blames-more-than-one-official-severe-punishment-to-be-meted-out-4961396.html

46 Matt Korda, "Flying Under The Radar: A Missile Accident in South Asia," *Federation of American Scientists*, April 4 2022, accessed December 16, 2022, https://fas.tghp.co.uk/publication/flying-under-the-radar-a-missile-accident-in-south-asia/

47 Vishal Thapar, "India 'Maximises' BrahMos Cruise Missile Range To Over 500 Km," *BW Businessworld,* January 2022, accessed December 16, 2022, http://businessworld.inhttps://www.businessworld.in/article/India-Maximises-BrahMos-Cruise-Missile-Range-To-Over-500-km-/21-01-2022-418377

48 Snehesh Alex Philip, "IAF to Increase Sukhois Armed with BrahMos Supersonic Cruise Missile with over 500km-Range," *ThePrint*, October 5, 2022, accessed December 16, 2022, https://theprint.in/defence/iaf-to-increase-sukhois-armed-with-brahmos-supersonic-cruise-missile-with-over-500km-range/1155562/

49 Vishal Thapar, "India 'Maximises' BrahMos Cruise Missile Range To Over 500 Km," *BW Businessworld,* January 2022, accessed December 16 2022. http://businessworld.inhttps://www.businessworld.in/article/India-Maximises-BrahMos-Cruise-Missile-Range-To-Over-500-km-/21-01-2022-418377 Snehesh Alex Philip, "IAF to Increase Sukhois Armed with BrahMos Supersonic Cruise Missile with over 500km-Range," *ThePrint*, October 5, 2022, accessed December 16, 2022, https://theprint.in/defence/iaf-to-increase-sukhois-armed-with-brahmos-supersonic-cruise-missile-with-over-500km-range/1155562/

50 Amitabh Sinha, "Misfire of BrahMos Missile Not Cause for Any Specific Concern: IAEA," *The Indian Express,* November 14, 2022, accessed December 17, 2022. https://indianexpress.com/article/world/misfire-of-brahmos-missile-not-cause-for-any-specific-concern-iaea-8266709/

FP Explainers, "India Accidentally Fires Missile into Pakistan: What and How Did It Happen?," *Firstpost,* March 12, 2022, accessed December 17, 2022, https://www.firstpost.com/india/india-accidentally-fires-missile-into-pakistan-what-and-how-did-it-happen-10452681.html

51 Matt Korda, "Flying Under The Radar: A Missile Accident in South Asia," *Federation of American Scientists*, April 4 2022, accessed December 17, 2022, https://fas.tghp.co.uk/publication/flying-under-the-radar-a-missile-accident-in-south-asia/

52 Huma Sidhiqui, "India Speeding up Development of Hypersonic BrahMos with Mach 5-7," *Financial Express*, May 2, 2023, accessed December 17, 2022, https://www.financialexpress.com/business/defence-india-speeding-up-development-of-hypersonic-brahmos-with-mach-5-7-3071445/

53 Snehesh Alex Philip, "Missile Accidentally Fired 'during Inspection' at Secret IAF Base, Pakistan 'Didn't Track,'" *ThePrint,* March 12, 2022, accessed December 17, 2022, https://theprint.in/defence/missile-accidentally-fired-during-inspection-at-secret-iaf-base-pakistan-didnt-track/870814/

54 Snehesh Alex Philip, "Missile Accidentally Fired 'during Inspection' at Secret IAF Base, Pakistan 'Didn't Track,'" *ThePrint,* March 12, 2022, accessed December 17, 2022, https://theprint.in/defence/missile-accidentally-fired-during-inspection-at-secret-iaf-base-pakistan-didnt-track/870814/

55 Snehesh Alex Philip, "Missile Accidentally Fired 'during Inspection' at Secret IAF Base, Pakistan 'Didn't Track,'" *ThePrint,* March 12, 2022, accessed December 17, 2022, https://theprint.in/defence/missile-accidentally-fired-during-inspection-at-secret-iaf-base-pakistan-didnt-track/870814/

56 Snehesh Alex Philip, "Missile Accidentally Fired 'during Inspection' at Secret IAF Base, Pakistan 'Didn't Track,'" *ThePrint,* March 12, 2022, accessed December 17, 2022, https://theprint.in/defence/missile-accidentally-fired-during-inspection-at-secret-iaf-base-pakistan-didnt-track/870814/

57 Snehesh Alex Philip, "Missile Accidentally Fired 'during Inspection' at Secret IAF Base, Pakistan 'Didn't Track,'" *ThePrint,* March 12, 2022, accessed December 17, 2022, https://theprint.in/defence/missile-accidentally-fired-during-inspection-at-secret-iaf-base-pakistan-didnt-track/870814/

58 Rajat Pandit, "Missile Firing: IAF Inquiry on, but Pak Seeks Joint Probe," *The Times of India*, March 13, 2022, accessed December 17, 2022, https://timesofindia.indiatimes.com/india/missile-firing-iaf-inquiry-on-but-pak-seeks-joint-probe/articleshow/90176509.cms?utm_source=contentofinterest&utm_medium=text&utm_campaign=cppst

59 Rajat Pandit, "Missile Firing: IAF Inquiry on, but Pak Seeks Joint Probe," *The Times of India*, March 13, 2022, accessed December 17, 2022, https://timesofindia.indiatimes.com/india/missile-firing-iaf-inquiry-on-but-pak-seeks-joint-probe/articleshow/90176509.cms?utm_source=contentofinterest&utm_medium=text&utm_campaign=cppst

60 Moeed Yusuf, "On March 9th, a supersonic projectile from India...," *X (formerly Twitter)*, March 11, 2022, accessed December 17, 2022, https://twitter.com/YusufMoeed/status/1502313842352795652

"NSA Yusuf Questions India's Ability to Handle Sensitive Technology after Missile Accident," *DAWN.COM*, March 11, 2022, accessed December 17, 2022, https://www.dawn.com/news/1679447

61 "NSA Yusuf Questions India's Ability to Handle Sensitive Technology after Missile Accident," *DAWN.COM*, March 11, 2022, accessed December 17, 2022, https://www.dawn.com/news/1679447

News Desk, "FM Qureshi Seeks Clarification from India on 'Flying Object,'" *The Express Tribune*, March 11, 2022, accessed December 17, 2022, https://tribune.com.pk/story/2347445/fm-qureshi-seeks-clarification-from-india-on-flying-object

62 Rahul Bedi, "India Needs to Come Clean on 'accidental Firing of Missile' to Avert Conflict with Pakistan," *The Wire*, March 12, 2022, accessed December 18, 2022, https://thewire.in/security/india-needs-to-come-clean-on-accidental-firing-of-missile-to-avert-conflict-with-pakistan

63 Mofa, "Indian Official Acceptance of Accidental Firing of Missile," *Ministry of Foreign Affairs, Pakistan*, March 12, 2022, accessed December 18, 2022, https://mofa.gov.pk/indian-official-acceptance-of-accidental-firing-of-missile/

64 Mofa, "Indian Official Acceptance of Accidental Firing of Missile," *Ministry of Foreign Affairs, Pakistan*, March 12, 2022, accessed December 18, 2022, https://mofa.gov.pk/indian-official-acceptance-of-accidental-firing-of-missile/

65 Mofa, "Indian Official Acceptance of Accidental Firing of Missile," *Ministry of Foreign Affairs, Pakistan*, March 12, 2022, accessed December 18, 2022, https://mofa.gov.pk/indian-official-acceptance-of-accidental-firing-of-missile/

66 Mofa, "Indian Official Acceptance of 'Accidental Firing' of Missile," *Ministry of Foreign Affairs*, March 12, 2022, accessed December 18, 2022, https://mofa.gov.pk/indian-official-acceptance-of-accidental-firing-of-missile-2/

67 Bureau, "Could Have Responded to India after Its Missile Landed in Pakistan, but Observed Restraint: PM Imran," *The Telegraph*. March 13 2022, accessed December 18 2022. https://www.telegraphindia.com/world/could-have-responded-to-india-after-its-missile-landed-in-pakistan-but-observed-restraint-pm-imran/cid/1855819

68 News Desk, "No Indication Indian Missile Launch Not an Accident: US," *The Express Tribune*, March 15, 2022, accessed December 18, 2022, https://tribune.com.pk/story/2348056/no-indication-indian-missile-launch-anything-other-than-an-accident-us

69 Office Spokesperson, "Foreign Minister Qureshi's Telephone Conversation with German Foreign Minister," *Ministry of Foreign Affairs*, March 14, 2022, accessed December 18, 2022, https://mofa.gov.pk/foreign-minister-qureshis-telephone-conversation-with-german-foreign-minister/

Spokesperson, "Foreign Minister Speaks with the UN Secretary General," *Ministry of Foreign Affairs, Government of Pakistan,* March 14, 2022, accessed December 18, 2022, https://mofa.gov.pk/foreign-minister-speaks-with-the-un-secretary-general/

70 SansadTV Bureau, "Defence Minister Rajnath Singh's Statement on the 'Inadvertent Firing of Missile on 9th March, 2022,'" *Sansad TV,* March 15, 2022, accessed December 19, 2022, https://sansadtv.nic.in/loksabha/defence-minister-rajnath-singhs-statement-on-the-inadvertent-firing-of-missile-on-9th-march-2022

"Review of Proceedings in Parliament," *All India Radio News,* March 15, 2022, accessed December 18, 2022, https://newsonair.gov.in/Main-News-Details.aspx?id=437246

Rajnath Singh, "RM Rajnath Singh's Statement on the "Inadvertent Firing of Missile on 9th March, 2022," *RS. Sansad TV,* March 15, 2022, accessed December 19, 2022, https://www.youtube.com/watch?v=1Rc3D9r92Bc

71 Rahul Bedi, "Did Rajnath Singh Subtly Suggest 'human Error' Led to Accidental Missile Firing?" *The Wire,* March 17, 2022, accessed December 19, 2022, https://thewire.in/security/india-accidental-missile-firing-human-error-rajnath-singh

72 PIB, "Statement on Accidental Firing of Missile," *Ministry of Defence,* March 11, 2022, accessed December 19, 2022, https://pib.gov.in/pib.gov.in/Pressreleaseshare.aspx?PRID=1805148

Rahul Bedi, "Did Rajnath Singh Subtly Suggest 'human Error' Led to Accidental Missile Firing?" *The Wire,* March 17, 2022, accessed December 19, 2022, https://thewire.in/security/india-accidental-missile-firing-human-error-rajnath-singh

73 Rajat Pandit, "India's Missile Systems Secure, Reliable: Rajnath," *The Times of India*, March 16, 2022, accessed December 19, 2022, https://timesofindia.indiatimes.com/india/indias-missile-systems-secure-reliable-rajnath/articleshow/90246711.cms?utm_source=contentofinterest&utm_medium=text&utm_campaign=cppst

Rahul Bedi, "After Missile Fiasco, Indian Officials Hope Spin Will Deflect Awkward Questions," *The Wire,* March 15, 2022, accessed December 19, 2022, https://thewire.in/government/missile-fiasco-india-pakistan-awkward-questions-spin

74 PTI, "Pakistan Rejects India's Response to Accidental Missile Launch; Demands Joint Probe," *The Times of India*, March 15, 2022, accessed December 19, 2022, https://timesofindia.indiatimes.com/india/pakistan-rejects-indias-response-to-accidental-missile-launch-demands-joint-probe/articleshow/90225544.cms

75 PTI, "Pakistan Rejects India's Response to Accidental Missile Launch; Demands Joint Probe," *The Economic Times,* March 15, 2022, accessed December 19, 2022, https://economictimes.indiatimes.com/news/defence/pakistan-rejects-indias-response-to-accidental-missile-launch-demands-joint-probe/articleshow/90225985.cms?utm_source=contentofinterest&utm_medium=text&utm_campaign=cppst

76 "X (formerly Twitter)," India has again claimed that the launch of its missile..., March 15, 2022, accessed December 19, 2022, https://twitter.com/YusufMoeed/status/1503711769138716672

77 Vishnu Som and Deepshikha Ghosh, "Indian Missile Accident Nearly Led to Pak Retaliation: Report," *NDTV.com*, March 16, 2022, accessed December 20, 2022, https://www.ndtv.com/india-news/indian-missile-accident-nearly-led-to-pak-retaliation-report-2825860

Agencies, "Pak Nearly Launched Retaliatory Strike after India's Accidental Missile Launch," *Livemint*, March 16, 2022, accessed December 20, 2022, https://www.livemint.com/news/world/pakistan-nearly-launched-retaliatory-strike-in-response-to-india-s-accidental-missile-launch-11647424134509.html

78 "248th Corps Commanders' Conference was held at GHQ. General Qamar Javed Bajwa, Chief of Army Staff (COAS) presided," *ISPR*, March 15 2022, accessed December 20, 2022, https://ispr.gov.pk/press-release-detail.php?id=6374

79 Ajay Banerjee, "Ambala IAF Officers to Join 'Accidental' Missile Firing Probe," *The Tribune,* March 15, 2022, https://www.tribuneindia.com/news/nation/ambala-iaf-officers-to-join-missile-probe-378114

80 Ajay Banerjee, "Ambala IAF Officers to Join 'Accidental' Missile Firing Probe," *The Tribune,* March 15, 2022, https://www.tribuneindia.com/news/nation/ambala-iaf-officers-to-join-missile-probe-378114

81 Ajay Banerjee, "Ambala IAF Officers to Join 'Accidental' Missile Firing Probe," *The Tribune,* March 15, 2022, https://www.tribuneindia.com/news/nation/ambala-iaf-officers-to-join-missile-probe-378114

82 MOFA, "Transcript of the Press Briefing by Spokesperson on Friday, 18 March, 2022," *Ministry of Foreign Affairs*, March 18, 2022, accessed December 20, 2022, https://mofa.gov.pk/transcript-of-the-press-briefing-by-spokesperson-on-friday-18-march-2022/

83 Office Spokesperson, "Foreign Minister Meets His Egyptian Counterpart," *Ministry of Foreign Affairs*, March 20, 2022, accessed December 20, 2022, https://mofa.gov.pk/foreign-minister-meets-his-egyptian-counterpart/

84 MOFA, "Foreign Minister Qureshi Received State Councilor and Foreign Minister of China," *Ministry of Foreign Affairs*, March 21, 2022, accessed December 2022, https://mofa.gov.pk/foreign-minister-qureshi-received-state-councilor-and-foreign-minister-of-china/

85 MOFA, "U.S. Under Secretary of State Uzra Zeya Calls on Foreign Minister," *Ministry of Foreign Affairs*, March 21, 2022, accessed December 20, 2022, https://mofa.gov.pk/u-s-under-secretary-of-state-uzra-zeya-calls-on-foreign-minister/

86 Spokesperson Office, "OIC Endorses Pakistan's Call for a Joint Probe into the 9 March 2022 Missile Incident," *Ministry of Foreign Affairs*, March 23, 2022, accessed December 20, 2022, https://mofa.gov.pk/oic-endorses-pakistans-call-for-a-joint-probe-into-the-9-march-2022-missile-incident/

PTI, "OIC Supports Pak's Call for Joint Probe in Accidental Firing of Projectile," *Business Standard*, March 24, 2022, accessed December 20, 2022, https://www.business-standard.com/article/international/oic-supports-pak-s-call-for-joint-probe-in-accidental-firing-of-projectile-122032400685_1.html

OIC, "Islamabad Declaration 48thSession of OIC Council of Foreign Ministers 'Partnering for Unity, Justice and Development'," *Organization of Islamic Cooperation*, March 23, 2022, accessed December 20, 2022, https://www.oic-oci.org/topic/ampg.asp?t_id=33947&t_ref=22694&lan=en

87 Yusuf Moeed, "#OICInPakistan Islamabad declaration #OIC48CFM highlights the threat to the region's security," *X* (formerly twitter), March 24, 2022, accessed December 22, 2022, https://twitter.com/YusufMoeed/status/1506955388792549381

88 Rahul Bedi, "Did Rajnath Singh Subtly Suggest 'human Error' Led to Accidental Missile Firing?" *The Wire*, March 17, 2022, accessed December 22, 2022, https://thewire.in/security/india-accidental-missile-firing-human-error-rajnath-singh

89 Tribune News Service, "4 IAF Officials under Lens Over Missile 'Misfire,'" *The Tribune*, March 24, 2022, accessed December 22 2022 https://www.tribunein-dia.com/news/nation/4-iaf-officials-under-lens-over-missile-misfire-380430

90 Pradip Sagar, "'Immature Handling' of Missile System Led to BrahMos Misfiring?" *The Week,* March 23, 2022, accessed December 22, 2022, https://www.theweek.in/news/india/2022/03/23/immature-handling-missile-system-led-to-brahmos-misfiring.html

91 News Desk, "Pak Missile Misfire: IAF Probe Blames More Than One Official; 'Severe Punishment' to Be Meted Out," *News18,* April 10, 2022, accessed December 22, 2022, https://www.news18.com/news/india/pak-missile-misfire-iaf-probe-blames-more-than-one-official-severe-punishment-to-be-meted-out-4961396.html

92 Ajit K. Dubey, "Missile Firing Incident: Probe Pins Blame on Group Captain," *Rediff,* March 23, 2022, accessed December 22, 2022, https://www.rediff.com/news/report/missile-firing-incident-probe-pins-blame-on-group-captain/20220323.htm

Pioneer News Service, "'IAF Conducting Court of Inquiry into Accidental Brahmos Firing,'" *The Pioneer,* March 24, 2022, accessed December 22, 2022, https://www.dailypioneer.com/2022/india/---iaf-conducting-court-of-inquiry-into-accidental-brahmos-firing---.html

93 TNN, "Lens on Group Captain for Missile Misfire," *The Times of India*, March 24, 2022, accessed December 22, 2022, https://timesofindia.indiatimes.com/india/lens-on-group-captain-for-missile-misfire/articleshow/90406580.cms?utm_source=contentofinterest&utm_medium=text&utm_campaign=cppst

94 TNN, "Lens on Group Captain for Missile Misfire," *The Times of India*, March 24, 2022, accessed December 22, 2022, https://timesofindia.indiatimes.com/india/lens-on-group-captain-for-missile-misfire/articleshow/90406580.cms?utm_source=contentofinterest&utm_medium=text&utm_campaign=cppst

95 Rahul Singh, "Accidental Missile Firing Probe Points to 'Human Error' as Likely Cause," *Hindustan Times*, March 23, 2022, accessed December 23, 2022, https://www.hindustantimes.com/india-news/accidental-missile-firing-probe-points-to-human-error-as-likely-cause-101648058869636.html

96 Vishal Thapar, "Commanding Officer, Others To Blame For Accidentally Firing BrahMos Missile Into Pakistan: Reports," *BW Businessworld*, April 11, 2022, accessed December 23, 2022, http://businessworld.inhttps://www.businessworld.in/article/Commanding-Officer-Others-To-Blame-For-Accidentally-Firing-BrahMos-Missile-Into-Pakistan-Reports/11-04-2022-425201.

97 TNN, "Lens on Group Captain for Missile Misfire," *The Times of India*, March 24, 2022, accessed December 22, 2022, https://timesofindia.indiatimes.com/india/lens-on-group-captain-for-missile-misfire/articleshow/90406580.cms?utm_source=contentofinterest&utm_medium=text&utm_campaign=cppst

98 News Desk, "No Technical Issue, Replies India as Philippines Seeks BrahMos Assurance Over 'Accidental Firing' Into Pak," *News18,* April 6, 2022, accessed December 22, 2022, https://www.news18.com/news/india/no-technical-issue-replies-india-as-philippines-seeks-brahmos-assurance-over-accidental-firing-into-pak-4946882.html

99 Business Today Desk, "Pakistan Missile Misfire: More than One Official Responsible, Says IAF," *Business Today,* April 11, 2022, accessed December 23, 2022, https://www.businesstoday.in/latest/story/pakistan-missile-misfire-more-than-one-official-responsible-says-iaf-329338-2022-04-11

100 Business Today Desk, "Pakistan Missile Misfire: More than One Official Responsible, Says IAF," *Business Today,* April 11, 2022, accessed December

23, 2022, https://www.businesstoday.in/latest/story/pakistan-missile-misfire-more-than-one-official-responsible-says-iaf-329338-2022-04-11

101 News Desk, "Pak Missile Misfire: IAF Probe Blames More Than One Official; 'Severe Punishment' to Be Meted Out," *News18*, April 10, 2022, accessed December 23, 2022, https://www.news18.com/news/india/pak-missile-misfire-iaf-probe-blames-more-than-one-official-severe-punishment-to-be-meted-out-4961396.html

Manjeet Negi, "More than One IAF Official Responsible in Accidental Missile Firing, Action to Be Taken Soon," India Today, April 11 2022, accessed December 23, 2022, https://www.indiatoday.in/india/story/more-than-one-iaf-official-responsible-in-accidental-missile-firing-action-to-be-taken-soon-1935809-2022-04-10

102 Vishal Thapar, "Commanding Officer, Others To Blame For Accidentally Firing BrahMos Missile Into Pakistan: Reports," *BW Businessworld,* April 11, 2022, accessed December 23, 2022, http://businessworld.inhttps://www.businessworld.in/article/Commanding-Officer-Others-To-Blame-For-Accidentally-Firing-BrahMos-Missile-Into-Pakistan-Reports/11-04-2022-425201

103 Ajit Dubey, "IAF Inquiry Blames More than One Official for BrahMos Accidental Firing, Swift, Severe Punishment to Be Awarded," *ThePrint,* April 10, 2022, accessed December 23, 2022, https://theprint.in/india/iaf-inquiry-blames-more-than-one-official-for-brahmos-accidental-firing-swift-severe-punishment-to-be-awarded/910825/

104 Tribune News Service, "IAF Sacks 3 Officers for Accidental Firing of Brahmos Missile That Landed in Pakistan," *The Tribune,* August 23, 2022, accessed December 23, 2022, https://www.tribuneindia.com/news/nation/iaf-sacks-3-officers-for-accidental-missile-fire-424654

Outlook Web Desk, "Accidental Missile Fire: Three Officers Terminated from Service, Enquiry Finds Deviation from Norms," *Outlook*, August 23, 2022, accessed December 24, 2022, https://www.outlookindia.com/national/accidental-missile-fire-three-officers-terminated-from-service-enquiry-finds-deviation-from-norms-news-218411

105 Tribune News Service, "IAF Sacks 3 Officers for Accidental Firing of Brahmos Missile That Landed in Pakistan," *The Tribune,* August 23, 2022, accessed December 23, 2022, https://www.tribuneindia.com/news/nation/iaf-sacks-3-officers-for-accidental-missile-fire-424654

106 Business Today Desk, "BrahMos Missile Misfire: IAF Officers Did Not Follow the SOPs, Says DG Rane," *Business Today,* August 26, 2022, accessed December 26, 2022, https://www.businesstoday.in/india-at-100/story/brahmos-missile-misfire-iaf-officers-did-not-follow-the-sops-says-dg-rane-345712-2022-08-26

107 "Pakistan Rejects India's Purported Closure of the Incident of the Firing of a Supersonic Missile into Pakistani Territory on 9 March, 2022 and Reiterates Demand for Joint Probe," *Ministry of Foreign Affairs,* August 24, 2022, accessed December 25, 2022, https://mofa.gov.pk/pakistan-rejects-indias-purported-closure-of-the-incident-of-the-firing-of-a-supersonic-missile-into-pakistani-territory-on-9-march-2022-and-reiterates-demand-for-joint-probe/

Mayank Singh, "Pakistan Rejects India's Closure Report of BrahMos Accidental Missile Firing," *The New Indian Express*, August 25 2022, accessed December 25, 2022, https://www.newindianexpress.com/nation/2022/aug/25/pakistan-rejects-indias-closure-report-of-brahmos-accidental-missile-firing-2491026.html

108 Nakshab Khan, "Brahmos Misfiring into Pakistan: Wing Commander Abhinav Sharma Files Case against Defence Ministry and IAF Chief," *Times Now,* March 14, 2023, accessed September 5, 2023, https://www.timesnownews.com

/india/brahmos-misfiring-into-pakistan-wing-commander-abhinav-sharma-files-case-against-defence-ministry-and-iaf-chief-article-98638952.

The Wire Staff, "Misfired Brahmos Which Landed in Pakistan Could Have Led to Warlike Situation: Union Govt to HC," *The Wire,* March 15, 2023, accessed September 3, 2023, https://thewire.in/security/misfired-brahmos-which-landed-in-pakistan-could-have-led-to-warlike-situation-union-govt-to-hc

2

MEDIA COVERAGE

Introduction

The media in India and Pakistan played an important role in reporting on events surrounding the missile accident and in analyzing its implications for bilateral relations and regional security. Chapter 1 focused narrowly on the media's role in amplifying the Indian and Pakistani states' narratives of events through anonymous leaks. However, the coverage of the crisis by the Indian and the Pakistani media also included articles informing the public about myriad issues and events including the DG-ISPR's press conference, details of the accident, the technical details of the BrahMos missile, capabilities of Pakistani air defense systems, the Indian Defence Ministry's official statement and the Indian Court of Inquiry etc.

The media's coverage of the aforementioned consisted of editorials, opinion pieces, televised debates with veteran journalists, long articles, X threads (previously Twitter) and YouTube channels of well-known journalists and editors. Chapter 2 focuses on the role of the Indian and the Pakistani media in reporting on and analyzing the missile accident. It looks at both Indian and Pakistani English-language print, digital and electronic media sources. Views of the Indian and Pakistani strategic communities, carried in the media are dealt with separately in Chapter 3 on the role of the two strategic communities.

The Indian Media's Reaction—Errant Missile Likely BrahMos

Amongst the very first reactions of the Indian media on the accident was its carrying of reports in the Pakistani media of an airplane crash in Mian

DOI: 10.4324/9781003543909-3

Channu "triggering an explosion and massive fire" with no reported casualties and the plane "charred beyond recognition."[1] Once it became clear that the reason behind the explosion was an errant Indian missile, an "Explainer" article in *Firstpost* on 12 March tried to situate the accident in the historic context of mishaps involving India and Pakistan.

> In 1997, an Indian Air Force MiG-25 jet flew into Pakistani airspace while on a reconnaissance mission. According to a *Business Today* report, the aircraft broke the sound barrier, sending a powerful sonic boom over Islamabad before returning unchallenged to the Indian territory.[2]

The Hindu also drew a connection between the missile accident and the accidental shooting down of an Indian Mi-17V5 Indian helicopter in Budgam by a Spyder surface-to-air missile system sourced from Israel and operated by the IAF during the aerial dogfight of Indian and Pakistani fighter jets on the day after India's aerial strike on Balakot.[3]

Seema Guha, the Foreign Editor for *Outlook Magazine* linked the missile accident to warnings by former US President Bill Clinton regarding the escalatory potential in India-Pakistan relations. She also called for a thorough investigation into the accident by India and commended Pakistan for its restrained response.

> Ahead of his visit to South Asia in 2000, former US President Bill Clinton had referred to the Line of Control between India and Pakistan as "the most dangerous place in the world" and a potential flashpoint for a nuclear flare-up in the volatile region marked by decades of sabre-rattling on both sides...the incident on Wednesday, March 9, of an accidental misfiring of an Indian missile that landed in Pakistan's Mian Channu area drives home the point made decades back by Clinton. It was an unarmed missile and landed in private property and luckily did not cause any major damage. It is not known if the backchannel lines were active and whether Pakistan was informed through these channels. Islamabad had acted with restraint to what could have been taken as provocation from India....It is for New Delhi now to find out what exactly happened last Wednesday and make sure that maintenance of such deadly missile systems is done under much stricter supervision.[4]

The Indian media also began informed speculation that the misfired missile was the BrahMos, based partly on Pakistan's description of its technical characteristics. Dinakar Peri, Senior Associate Editor at *The Hindu* speculated in a 16 March article that

> the description given by Pakistan on the technical characteristics of the missile fits that of BrahMos, product of a joint venture between India and

> Russia and named after the Brahmaputra and Moskva rivers. BrahMos is a potent precision strike weapon with variants that can be launched from land, sea and air and a range of 290 kms deployed by all three Services. No similar capability exists in the region and intercepting a supersonic cruise missile which can also maneuver and fly low to evade detection is extremely difficult.[5]

He also echoed the general Indian media sentiment that the incident had brought to the fore the "need to review and make protocols foolproof to avoid any recurrence."[6]

Need for an Impartial Indian probe but Not a Joint Probe

Editorial opinion in India's major newspapers was uniformly in favor of an impartial Indian probe to ascertain the cause of the accident, affix responsibility and to negotiate various risk-reduction measures. In a frank editorial on 14 March, *The Hindu* warned of the serious escalatory potential of the accident and called on the Indian Court of Inquiry to "get to the truth of the missile malfunction" even as it deemed Pakistan's demand for a joint probe "unwarranted."

> The description by Pakistan—that the missile was travelling at three times the speed of sound, at 40,000 feet, and is a surface-to-surface missile—has led to speculation that the accident involves the BrahMos supersonic cruise missile which is now in the inventory of India's three Services. India became a member of the Missile Technology Control Regime in 2016, an acceptance by major powers of India's status as a reliable defence partner...India must leave no scope for any doubts about its capacity to handle nuclear and other military assets. That objective can be achieved without a joint probe with Pakistan or any international involvement, but the objective must be achieved nevertheless.[7]

A *Times of India* editorial on 13 March clearly identified the errant missile as a BrahMos. Describing the missile accident as a "truly scary event," the editorial brought up both the technical and human error possibilities and enjoined the IAF to conduct a thorough investigation. It also called on India and Pakistan to develop elaborate "risk-reduction protocols" and ended with a candid recommendation for those responsible to be held accountable.

> What is baffling here is that the accident happened with much-tested BrahMos, which has been in service for many years now. And the missile launch system comes with multiple mechanical and electronic locks. Therefore, either these locks were accidentally bypassed by operators

> during the training exercise or there was a serious technical malfunction....If human error is to blame, then the operational, command and control procedures must be reviewed. On the other hand, a technical glitch of this nature will cast doubts over the weapon system itself, especially at a time when India is looking to export BrahMos to countries like the Philippines. True, a court of inquiry has been ordered to determine the chain of events that led to the mishap. But responsibility needs to be fixed and heads must roll if required.[8]

An editorial in the *Indian Express* two days later praised Pakistan for acting with "maturity." It also enjoined Delhi to conduct a thorough probe and "fix responsibility." The editorial also speculated that the Pakistani delay in announcing and the Indian delay in admitting to the accident could have been a result of secret parleys between the two countries.

> Over the last few days, Pakistan and India have given the world an unexpected real time demonstration of how two nuclear armed neighbours should behave to prevent a crisis from forming and escalating...Pakistan has acted with maturity and responsibility. By all accounts, it seems that the delay on both sides in announcing and acknowledging the incident was also intentional, perhaps giving both sides crucial time to make contact through the available channels, not all of them public...At the same time, the incident cannot be brushed away...Apart from fast-tracking the court of inquiry and handing out quick and firm punishment to those responsible for this mistake, this incident shows that Delhi and Islamabad must, at the very least, resume talks on conventional and nuclear confidence building mechanisms that have not been held for many years.[9]

Writing in *The Wire* on 12 March, veteran journalist and commenter on strategic issues Rahul Bedi criticized New Delhi for its 24-hour delay following the Pakistani press conference in "conceding its colossal error" that had the potential of "escalating into a conflagration of apocalyptic proportions." Taking issue with the lack of transparency in New Delhi's initial response, Bedi called on the Indian government to ensure a high degree of accountability in its Court of Enquiry.

> That it (the missile) did not hurt anyone can be attributed to providence and the collective karma of both countries...The entire missile episode, however, is shrouded in thick fog in New Delhi, which has not only declined to identify the missile involved but also which of the three services fired it, all of which will now be investigated by the COI of which too no details were forthcoming...the CoI into the rogue missile incidence needs an exceptionally high degree of accountability.[10]

Bedi also castigated some sections of the media for their collusion with the establishment and for helping spread the government's version of events that was riddled with contradictory claims.

> Faced with an embarrassing narrative that the MoD's (Ministry of Defence) own conduct and statements have contributed to, ministry officials have recently launched "damage control" manoeuvres with the help of "obliging" journalists.... Quoting unnamed defence and security sources, some recent reports sought to contradict Pakistan's claims of the accidentally fired BrahMos missile having changed direction after flying south to enter Pakistan. Some of these correspondents, seemingly at the behest of senior government officials, claimed that the "practice" missile had, in actual fact, followed its "correct trajectory" that had been charted had it been in a conflict situation. These reports dismissed speculation that the missile had been headed for the Mahajan Field Firing Range in Rajasthan, before inexplicably changing course and entering Pakistan. These bewildering, and at times contradictory, accounts from "sources in the know" also claimed that the missile – exactly similar to the armed version minus its warhead – was the one that ended up being "accidentally" launched from a "secret IAF satellite station," one report stated. Why an Indian missile primed for combat would be programmed to hit a random wall in Pakistan is not explained. These accounts also claimed, apparently under official tutelage, that the "relevant Pakistani authorities" had been informed of the accidental launch, but declined to state who this was or, for that matter, the date and time this had ensued.[11]

Sushant Singh, Deputy Editor of *The Indian Express* with experience in international affairs and national security[12] contended in a 13 March article that the missile misfire had "jeopardized India's long-standing reputation in nuclear safety." Lauding the Pakistani military for responding with "great maturity," he called for a thorough review of "systems and processes."

> India and Pakistan are nuclear weapon states that came close to climbing the escalation ladder in the aftermath of the Balakot airstrikes just three years ago...The accidental firing of an Indian missile has brought the spotlight on those risks again...Pakistan has claimed that the DGMO-level hotline wasn't activated to inform it of the accidental firing of the missile and its trajectory. This has been neither disputed nor explained by the Indian side...Imagine if a missile had fallen on Indian territory and the Pakistani side had claimed "accidental firing", would the Indian political leadership and jingoistic media have accepted that explanation?...The lives of 1.6 billion people of India and Pakistan cannot be dependent on such luck breaks...A thorough review of systems and processes to avert

> another accident is an imperative and would also need to focus on the professionalism of the security establishment, a record blighted by political favoritism and jingoism in recent years.[13]

Technical Malfunction, Gaps in Pakistan's Air Defense, Need for Risk Reduction Measures

Alongside widespread calls by the Indian media for a transparent investigation and for fixing accountability, there were also articles that analyzed the missile's flight path deep into Pakistan as weaknesses in its air defense system. Abhinandan Mishra, analyzed in the *The Sunday Guardian Live* that

> the "accidental firing" of an unarmed BrahMos supersonic missile by India that was caused by a "technical malfunction" in "course of a routine maintenance" from a base operated by the Indian Air Force (IAF), has raised questions inside Pakistan's military and political circle as to how did the missile travel inside Pakistan air space for three minutes and 44 seconds without being intercepted by any of its air defence systems. During this time the missile covered 124 km before it crash-landed near the country's eastern city of Mian Channu. In March 2017, the Pakistan army inducted HQ-16 or LY-80, a Low to Medium Altitude Air Defence System (LOMADS) to safeguard its airspace from incoming missiles and fighter jets....Its effective range, as per claims made by Chinese military experts, in case of cruise missile is 3.5-18 km. Many Chinese military sites had listed HQ-16 as an effective deterrent against BrahMos....Mian Channu is located around 160 km from the town of Bahawalpur or around 2.30 minutes away at BrahMos speed and 8 minutes away from capital Islamabad. As per (Major General Babar) Iftikhar's submission, the Pakistan military and its defence systems were aware of the launch of this missile for the entire 7 minutes in which it was flying.[14]

News9live, a Noida-based Indian digital outlet, carried a 13 March article by Islamabad-based journalist Aarzoo Kazmi praising Pakistan's "restrained response" to the Indian missile accident which "averted grave consequences" and supporting Pakistan's demand for a joint probe. Kazmi's also pointed out that the Indian missile misfire was a violation of the provisions of the 2005 India-Pakistan Pre-notification of Flight Testing of Ballistic Missiles Agreement.

> In the light of that agreement (India-Pakistan's Pre-notification of Flight Testing of Ballistic Missiles Agreement-2005), there was irresponsible violation of the provisions by India, and Pakistan has the legitimate right to seek a joint probe for getting a transparent outcome....Now, some critics can judge Pakistan's response as timid, but on [the] other hand it can't

> be denied that Rawalpindi's restraint has arrested further deterioration of [the] situation. Some observers have seen it as a proficient step to compel India to retract from the probable opportunity of resorting to the so-called 'Cold Start' doctrine in current volatile geopolitical situation.[15]

Even at this early stage, some sections of the Indian media were already making efforts to repair the damage caused to the reputation of the BrahMos missile system and by extension, to the IAF and the Indian security establishment. A 14 March video by the *Hindustan Times* claimed that, "in a major boost to missile shakti," the "hypersonic" BrahMos missile's strike range would be extended to 800 km, enabling it to "soon hit deep inside China."[16] Senior news writer Sudeshna Singh, writing for the right-leaning *Republic World* on 14 March, accused China of provoking Pakistan into calling for a joint probe.

> Prone to meddling in India's affairs, China on Monday, March 14, provoked Pakistan to hold talks and seek a thorough probe into the accidental missile-firing incident. China's provocation comes even as Pakistan and India are already in constant touch in regards to the matter, and have been discussing and deliberating on the topic of enquiry.[17]

OpIndia also gave prominent play to allegations made on X (formerly Twitter) by a Pakistani security hawk and conspiracy-theorist Zaid Hamid that the missile attack was a deliberate attempt by India to probe Pakistani defenses and was a prelude to a real attack while also being a celebratory firing in the wake of Yogi Adityanath's victory in the Uttar Pradesh elections.

> On Thursday, March 10, the Pakistani Army claimed that a 'missile' from India was launched from Hisar, violated the Pakistani airspace and fell in Mian Channu city in Khanewal district of the Punjab province of Pakistan....While Imran Khan faces massive protests and unrest in his own country, the claims of the so-called "supersonic missile" travelling inside Pakistan has brought out some wild conspiracy theories from Pakistan. One Pakistani "expert" Zaid Hamid, who keeps on issuing threats against Hindus and spreads fake news against India, had claimed that the so-called supersonic missile was nuclear-capable, and it was an "attack by India"....Hamid claimed that the nuclear missile launch was India's plan to test and evaluate Pakistan's entire nuclear doctrine and validate their response. Hamid further claimed that India's "nuclear attack" was successful and next time, India will launch a cruise missile with a real warhead. After spinning a whole war scenario, Hamid then stated that he has been told by some of his friends that the "nuclear missile launch" was done to celebrate CM Yogi Adityanath's victory in the UP elections.[18]

Another article in the same publication on 11 March attempted to make light of the whole accident by highlighting how it had "tickled the funny bones of many social users" who took to social media websites to share "hilarious memes" of the incident.

> A social media user shared on Twitter a snippet from the popular Bollywood movie "Ek aur Ek Gyarah" wherein actors Sanjay Dutt and Govinda, perched inside a battle tank, indiscriminately press the buttons that lead to the firing of missiles and bullets at the villains. A Twitter user shared an image of PM Modi allegedly involved in some experimentation, to insinuate that the missile was fired by him and to poke fun at Pakistan.[19]

The *Hindustan Times* sought to downplay the seriousness of the incident by characterizing the BrahMos missile as a "tactical system"

> Pakistan has been trying to rake up the issue of accidental missile firing at the international stage without realising that the BrahMos missile is just a tactical system and was unarmed at the time of the incident. The missile also did not cause any major damage to property or any loss of life in the Mian Channu town of Pakistan. India has also regretted the incident.[20]

After an eloquent summary of nuclear and missile misfire accidents in history involving the US, Soviet Union, UK, France, Taiwan, Russia and Iran, Praveen Swami (Group Consulting Editor at Network18 Group) called on India and Pakistan to engage in a dialogue on risk-reduction measures in an article for *The Print*.

> Little imagination is needed, though, to see that in other circumstances—a military stand-off, ongoing airstrikes, even large-scale deaths in a mishap—things might go somewhat differently. Islamabad reacted to the Mian Channu incident with admirable sangfroid; in other circumstances, it or New Delhi might have acted differently....Islamabad and New Delhi must engage in the granular, meaningful conversation needed to learn from the Mian Channu experience, and apply its lessons to their nuclear weapons contexts.[21]

Criticism of India's Delayed Response, BrahMos Not "Nuclear-Capable"

The Indian media also gave prominent coverage to US support for the Indian position that the missile firing was purely accidental.[22] Substantial coverage was also given to an article in *Bloomberg*[23] which claimed that Pakistan had planned a retaliatory strike with a similar missile in response to the errant Indian missile but held back.[24]

In her primetime show "No Spin," left-leaning NDTV anchor and veteran journalist Nidhi Razdhan introduced the missile accident as "the craziest story in recent times." She further averred that "this was a very serious accident" as "India and Pakistan are two hostile nuclear neighbours." She also opined that the accident "raises serious questions about the safety systems we have in place as a responsible nuclear power." The panel for the discussion that followed included Suhasini Haider, National Editor of *The Hindu*. Suhasini worried that "there is a greater danger in the way in which the government has managed its public explanation over this," especially its explanation "coming 48 hours after the missile malfunction seems extremely, extremely tardy." Suhasini then castigated the Indian government for the language of its statement which "shows this as something to move on from." She also speculated that "there would have been an escalation of some sort had Pakistan detected an Indian missile coming into its airspace and landing there without the necessary follow-up call from India" and that at the "DGMO-level or higher-up, a call was made explaining what had happened which is the only reason why it took 24 hours for the Pakistani establishment to issue a statement."[25]

The panel also included veteran Pakistani journalist Munizae Zahangir (Founding Editor, VOICEPK.NET) who informed that the Pakistani government's preoccupation with its domestic political crisis in the form of a no-confidence motion faced by Prime Minister Imran Khan meant that "the focus of the government is not on foreign policy." On being asked whether there was criticism directed at the Pakistani military for failing to shoot down the incoming missile, she informed that the criticism was mostly directed at the Pakistani government by the opposition Pakistan People's Party (PPP) and the Pakistan Muslim League-Nawaz (PML-N).[26]

Veteran journalist Shekhar Gupta, in his popular "Cut the Clutter" show, characterized the BrahMos missile as a "chota (small) level hypersonic missile." Gupta ruled out the Pakistani claim that the missile had been originally targeted at the Mahajan Firing Range by pointing out that no Notice to Airmen (NOTAM) had been issued, as that airspace was busy with civilian flights on the evening of 9 March and that this was the reason why Indian missile tests usually occurred on its eastern coast, off Odisha. Gupta also ruled out that India had deliberately fired the missile into Pakistan as that airspace too was filled with civilian aircraft. Gupta confirmed that the missile had no warhead as its 200 kg explosives (carrying capacity) would have otherwise caused a lot more destruction including possibly human lives that would have increased pressure on Pakistan to respond.

He also lauded Pakistan for responding "with a great deal of maturity." Without providing specific evidence, Gupta claimed that he was "quite sure" and would be "damned" if it did not turn out that, "as soon as it (missile misfire) happened, and in whatever time it takes to convey the message to Delhi, the DGMO on the Indian side or somebody in his office got in touch with the counterpart on the Pakistani side and said that look, we have a loose missile, we didn't fire it, something went wrong, it doesn't have a warhead, please calm down, this is not an attack."[27]

A particularly revealing aspect in Gupta's analysis was his knowledge of the specifics of the incident, informing his audience that the missile launcher in question "had three missiles" and when, "one was launched…immediately the power was cutoff to the vehicle so others also do not get launched." Gupta also stressed that "contrary to what people believe, the BrahMos is not nuclear-capable" which is the reason why India can export it to countries like the Philippines in keeping with the MTCR.[28] Gupta ended by enjoining India and Pakistan to strengthen their protocols and to ensure sustained communication and engagement.

Writing in the *Indian Express,* journalist Krishn Kaushik characterized the missile misfire as "extremely rare" and attempted to explain both the protocol followed during the tests and the circumstances that could result in an accident. He also reckoned that India and Pakistan would have to notify each other of such tests in accordance with the Pre-notification of Flight Testing of Ballistic Missiles Agreement signed in 2005 (although the agreement is supposed to apply only to ballistic missiles and not cruise missiles).

> Before the test, the country must issue Notice to Air Missions (NOTAM) or Navigational Warning (NAVAREA) to alert aviation pilots and seafarers, respectively. Also, the testing country must ensure that the launch site is not within 40 km, and the planned impact area is not within 75 km of either the International Boundary (IB) or the Line of Control (LoC). The planned trajectory should not cross the IB or the LoC and must maintain a horizontal distance of at least 40 km from the border. The testing country must notify the other nation "no less than three days in advance of the commencement of a five day launch window within which it intends to undertake flight tests of any land or sea launched, surface-to-surface ballistic missile". The pre-notification has to be "conveyed through the respective Foreign Offices and the High Commissions, as per the format annexed to this Agreement."[29]

From Technical Error to Human Error

In a 17 March article for *The Wire*, veteran journalist Rahul Bedi analyzed that Defence Minister Rajnath Singh's carefully crafted statement to Parliament on 15 March hinted at the prospect of human error, moving away from the Defence Ministry's earlier statement blaming the accident on a "technical malfunction." He also summarized the Indian defense industry's perceptions of the rationale behind the government allowing for the human error angle to emerge in Singh's statement.

> Additional understated indicators in Singh's statement further support the emerging 'human error' narrative....Singh's time frame for the unintended missile launch is corroborated by the Pakistan Air Force (PAF) which claims to have closely monitored its take-off and flight path..... Industry officials also pointed out that the authorities were "commercially conscious" of the fact that any hint of a "technical malfunction" in the BrahMos triggering its chance flight could adversely impact the $375 million deal which Delhi signed with the Philippines in January for three of its land-attack batteries. They reasoned that such fears could also result in other potential BrahMos customers like Brazil, Indonesia, South Africa and Vietnam, amongst others, querying the missile's operational credentials.[30]

The Print also relayed an Asian News International (ANI) news service article on 18 March claiming a Pakistani failed missile test that crashed in Sindh.

> The residents of Jamshoro in Pakistan's Sindh province noticed an unidentified flying object falling from the sky with a tail of smoke on Thursday. The object clearly resembled a rocket or a missile. According to information gathered from social media, the object was a missile, fired by Pakistan from its test range at Sindh. The test was scheduled at 11 am which was postponed by an hour owing to a fault in the TEL (Transporter Erector Launcher). It was finally conducted at noon. However, seconds after the launch the missile was seen descending from the path, clearly falling short of its desired target and crashed near Thana Bula Khan in Sindh...According to social media accounts in Pakistan, the local administration, however, refuted any such claim saying that it was a regular Mortar Tracer Round that was fired from the nearby range. It is unlikely, however, that a mortar with a maximum range of 5 km will have a tracer projectile rise that high...Pakistani news agency the Conflict News Pakistan reported that Pakistan might have test-fired a missile in response to a previous missile mistakenly fired from India.[31]

Chennai-based journalist Syed Ali Mujtabal, opined in the *Siasat Daily* on 19 March that the Defence Ministry's terse statement and admission of the errant missile launch only occurred because of pressure from the Coalition for Nuclear Disarmament and Peace (CNDP), a national coalition of NGO's working towards nuclear disarmament.

It was only about 48 hours after the mis-launch that the Indian Defence Ministry, through a press statement, chose to confirm the incident, describing the incident as accidental and terming it "deeply regrettable." This "deeply regrettable" word could have been blatantly denied had the Coalition for Nuclear Disarmament and Peace (CNDP), group not expressed its grave concern over the accidental launch of the BrahMos supersonic missile by India....It (CNDP) has also sought an explanation from India about its failure to alert the Pakistani Government immediately after the mis-launch. This was a deliberate violation of agreed norms, making India a suspect nation that cannot handle its nuclear weapons. The CNDP has also urged the government of India to immediately work out credible and effective confidence-building measures (CBMs) with Pakistan, revive the multiple channels of communication with them to avoid any misunderstanding.[32]

Court of Inquiry Finds Officer "Blameworthy"

The Indian media's focus continued to be on ongoing developments surrounding the Indian Court of Inquiry and several media outlets carried the news that "prima facie, a Group Captain-rank officer is being seen as blameworthy."[33] *The Print* explained further

> An Air Vice Marshal from the Air Force headquarters is carrying out a detailed investigation into the accidental firing of the tactical BrahMos supersonic cruise missile which landed in Pakistan on March 9. Government sources told ANI that the detailed probe by the Air Force officer (equivalent to Major General in the Army) is still going on but prima facie, a Group Captain-rank officer is being seen as blameworthy. The officer was in charge of the mobile command post of the missile system when the accidental firing happened during a Command Air Staff Inspection (CASI) at its home base....Pakistan has been trying to rake up the issue of accidental missile firing at the international stage without realising that the BrahMos missile is just a tactical system and was unarmed at the time of the incident.[34]

Putting a positive spin on events, a rather creative video titled "Pakistan's Lies Busted As Investigation Reveals 'Human Error'," on the website of the pro-establishment, right-leaning *TimesNow* news channel informed that

> the probe has revealed that the missile was fired due to a human error and not willingly as alleged by Pakistan. Group captain and certain other officials are under the lenses of the inquiry panel for these alleged lapses and the investigation is still underway. The report has made it clear that

> the claims of Pakistan were not factual. Pakistan had been claiming that the fire of BrahMos was not by accident but was designed by India. Meanwhile, the final results of the probe are yet to be announced by the panel.[35]

In early April, *TimesNow* reported that the Court of Inquiry was "reportedly probing four officers, including the commanding officer of the unit from which the BrahMos missile was 'accidentally' misfired" and that it has "held the BrahMos unit Commanding Officer, a Group Captain ranking officer and others responsible for the 'human' error."[36] A 11 April article in the *Times of India* also confirmed that the inquiry into the missile accident had been completed and the perpetrators identified. The article also explained the procedural aspects of the Indian military legal system.

> The Court of Inquiry (COI) has held the CO, a Group Captain-rank officer (Colonel) and the others prima facie responsible for the unprecedented incident, sources said on Sunday. The inquiry has been completed after examining the entire sequence of events as well as the acts of omission and commission....The final COI findings are now being "legally vetted" before the report is submitted to higher authorities for further action. "The lapses and extent of culpability of those in command has to be established to ensure the changes stand up legally," a source said. In the Indian military legal system, once the COI report is approved, the summary of evidence is recorded and then followed by a court martial or the actual trial. Disciplinary action can range from a severe reprimand and loss of seniority to even summary dismissal from service for such lapses.[37]

Missile Launch Deliberate

Abhishek Srivastava, a content editor at *NewsGram.com* reckoned in a 27 April article that India's claimed accidental launch of a missile into Pakistan was in fact a deliberate (and ultimately successful) move to test Pakistan's operational readiness.

> Though India has a history of military accidents, friendly fires or even SOP failures, the recent BrahMos missile misfire seems to be intentional and planned rather than an accidental one....Some of the newest and most advanced Chinese air defence systems that are currently the backbone of Pakistan's air defence are: Low-to-Medium Altitude Air Defence System: HQ-16AE/LY-80, Short-Range Air Defence system: HQ-7B/FM-90 and High-to-Medium Altitude Air Defence System: HQ-9/P. As Chinese military equipment suffers extreme incompatibility issues with western

> or European equipment, Pakistan had to also buy several Chinese radar and tracking systems like the IBIS-150 radars. Pakistan's acceptance of intrusion by an Indian missile into its air space has exposed the reality of the Chinese air defence systems that they purchased spending billions of dollars….Pakistan's Chinese radars were able to track the missile, but still didn't fire any retaliatory missile to destroy the Indian missile while it was in the air….Here Pakistan's air defence command stands exposed. Either they lied about being able to track the Indian missile or else they feared destroying an Indian missile intruding their air space…..If we put aside India's acceptance that it was an accidental misfire, it seems that India successfully tested Pakistan's operational readiness. The incident intensifies the popular belief that Pakistan won't be able to detect or counter any future missile attack from India that involves the supersonic BrahMosor its upcoming hyper-sonic variant BrahMos 2.[38]

A 29 August article in *Editorji* (and an accompanying video on its website),[39]echoed Srivastava's contention and pointed out that India's errant missile had exposed the weakness of the Pakistani air defense system, and by extension that of China, its key supplier.

> The incident has exposed the Pakistani military's dependence on Chinese weapons, and how those weapons might be ineffective against India's deadly arsenal…..One of these (Pakistani air defence systems) is the HQ-16 LOMADS, or Low to Medium Altitude Air Defence System. It is made by China, and was inducted into Pakistani forces in 2017. Its effective range is claimed to be 3.5-18 km against cruise missiles, like BrahMos. It can intercept low-flying targets at up to 40 km. The HQ-16 is claimed to have the capability to hit targets at 400 to 10,000 metre-altitude. Chinese commentators have called it a BrahMos deterrent. The second is the HQ-9 HIMADS, or High to Medium Altitude Air Defence System. This is also made by China, and was inducted into Pakistani forces in 2021. It is claimed to have the capability to intercept multiple aerial targets like cruise missiles, and aircraft. Pakistan military claims it can also intercept beyond-visual-range, or BVR, weapons at a range of 100 km. The HQ-9 has a single-shot kill probability, along with a 120 km detection range, and 90 km tracking range. But when the BrahMos missile flew over 100 km inside Pakistani territory, these defence systems were conspicuously absent. Why wasn't the rogue missile shot down?[40]

The aforementioned video analyzed four possibilities for Pakistan's non-response to the missile intrusion: 1) "the Pakistani military was caught off-guard and unprepared," 2) "Pakistani missile defense systems simply lacked the capability to intercept the BrahMos," 3) "a missile defense system was not in range of the rogue missile," and 4) "Islamabad, the political capital and Rawalpindi, the military centre, did not want the situation to escalate" as it would expose Pakistan's own inability to defend its airspace.[41]

Major (Retd) Gaurav Arya, consulting editor for strategic affairs with the right-leaning *Republic TV*, also opined in an interview with Pakistani journalist Arzoo Kazmi that the Indian firing of the BrahMos missile into Pakistan was a deliberate move and that a weapon as sophisticated as the BrahMos could not have been launched accidentally.[42]

Human Error Established—Three Officers Sacked

Although articles in the media had claimed the completion of the Court of Enquiry by early April, further details regarding the accused and the punishment meted out to them only emerged in late August of 2022. The Indian media reported on the conclusions of the Court of Inquiry, made public in a statement by the IAF that it had sacked three officers for deviation from Standard Operating Procedures (SOPs) resulting in the accidental firing.

> CLOSURE OF THE INCIDENT OF ACCIDENTAL BRAHMOS MISSILE FIRING ON 09 MARCH 2022
>
> A BrahMosmissile was accidentally fired on 09 March 2022. A Court of Inquiry (CoI), set up to establish the facts of the case, including fixing responsibility for the incident, found that deviation from the Standard Operating Procedures (SOP) by three officers led to the accidental firing of the missile. These three officers have primarily been held responsible for the incident. Their services have been terminated by the Central Government with immediate effect. Termination orders have been served upon the officers on 23rd Aug 2022."[43]

Despite the IAF statement's brevity, the Indian media was quick to notice the stark difference in the Indian probe's conclusive finding of human error being the cause of the accident as opposed to the Defence Ministry's initial claim of a technical malfunction. A 24 August article in *The Tribune* also claimed that the CoI had initially "looked into the role of at least four IAF officials."

> The services of a Group Captain and two Wing Commanders have been terminated with immediate effect....The Air Force said the three officers had "primarily been held responsible for the incident". The initial probe indicated a "human error", and not a technical fault, had led to the fire. The CoI found the Commanding Officer (Group Captain) and others "blameworthy of serious operational lapse". The IAF decided to

> terminate the services of the three officers as the matter was seen a serious lapse capable of triggering a war between India and Pakistan.[44]

A day earlier, the *Deccan Herald* also highlighted the shift in claimed rationale from technical malfunction to human error.

> The Centre on Tuesday sacked three senior Indian Air Force (IAF) officers who were responsible for an accidental firing of a BrahMos missile that landed in Pakistan earlier this year, causing huge embarrassment to India which takes pride in the safety and security of its arsenal....Two days after the incident, a Defence Ministry spokesperson had stated that a "technical malfunction in the course of a routine maintenance" led to the accidental firing.[45]

An editorial in the same newspaper on 2 September welcomed the outcome of the CoI and commended Pakistan for its mature response. It depicted the dismissal of the three officers as an honest Indian attempt at fixing accountability and demoted the missile accident as an insignificant issue in bilateral relations. It ended by urging the Indian government to plug any remaining gaps in technical systems and operational protocols for firing of missiles.

> The action taken against three IAF officers for firing a BrahMos missile into Pakistani territory earlier this year is welcome, and making the action public is especially so....It was lucky that Pakistan did not respond to it in kind....There was no reason to judge an incoming missile as an accident in the none-too-friendly atmosphere that exists between India and Pakistan. Though the Pakistani officer/s who decided that it was a mistake did well to do so, it would be no surprise if they had to answer some tough questions from the top. It is a moot question whether they followed the Standard Operating Procedure (SOP) prescribed for such situations.....The admission of its mistake by India and the action against the officers should allay Pakistan's concerns and apprehensions....Since the matter is over and the officers have been punished in an exemplary manner, it should no longer be an issue in bilateral relations. But there are some questions that India cannot ignore....They are about whether there is a 101% fool-proof system that rules out a technical or human error and whether the system has ensured against anyone going rogue or deviant for any reason.[46]

Pakistani Media Coverage—Focus on India's Regret

The Pakistani media's coverage of the missile incident initially focused on the Pakistani government's 10 March press conference by the DG-ISPR that a "supersonic Indian projectile without a warhead" had intruded into Pakistan's airspace and how the press conference had laid to rest earlier media reports of a plane crash in Mian Channu.

> Earlier, conflicting reports from Mian Channu were received Wednesday night, with some people claiming that an aircraft had crashed near the city and others saying that the object did not look like an aircraft. AVM (Air Vice Marshall) Zia said that the initial investigation suggested that it was an unarmed supersonic missile. Gen Iftikhar refused to comment on the possible motives of New Delhi and said it was for India to explain the development....Journalists at the press conference wanted to know if India had tried to test Pakistan's defence capability. The ISPR-DG said that India has not communicated anything about the object through DGMOs hotline. However, some people have claimed that the flying object was a variant of BrahMos missile....Social media posts showed what looks like a cowling of missile or small engine.[47]

The Pakistani media also focused on the Pakistani Foreign Office's subsequent summoning of the Indian Charge d'Affaires to lodge a protest.[48] *The Dawn* was careful to highlight that India "deeply regrets" the accidental firing of the missile[49] and took aim at the generic brevity of the Indian Defence Ministry's statement. It also speculated that Pakistan's relatively muted response was a result of the ongoing back-channel talks between the two countries.

> The Indian Defence Ministry's three-line statement explaining the missile fire came a day after Pakistan demanded to know the details of the incident....The statement gave no details of the nature of the missile, or even that it was unarmed, as noted by the military's spokesman, but expressed relief no one was harmed....The *Indian Express* newspaper offered sketchy details of the incident, but did not name its source. It described the projectile as an "unarmed supersonic Indian missile" that landed "124km deep inside Pakistan Wednesday". That the apparent accident took place in the domain of a nuclear-armed India, and impacted the territory of its nuclear archrival Pakistan would, of course, be studied by defence analysts for its fraught possibilities. However, it has happened amidst rumours of back-channel diplomatic talks between the two countries, which could explain it wasn't going beyond a reprimand and demand for explanation.[50]

Dunya News also emphasized that India had admitted its "mistake" in the missile accident and worried about the implications of any future accident for regional security.[51] The media focused on the Pakistani Foreign Office's protest over the Indian delay in acknowledging the accident, its re-summoning of the Indian Charge d'Affaires,[52] and its demand for answers to its seven key questions to India as well as calls for a joint probe.[53] Coverage was also given to the Pakistani National Security Adviser Moeef Yusuf's 11 March statements questioning India's capacity to safely handle sensitive technology[54] and his demand for an investigation into the misfire.[55] Taking note of the official US position, the Pakistani media carried the US State Department's muted reaction to the incident, including its support for the Indian explanation that the missile intrusion was nothing more than an accident.[56]

Technical Malfunction—India Not a Responsible State, Opportunity to Reverse-Engineer

Even as it provided real-time coverage of the official reactions in the immediate aftermath of the missile accident, sections of the Pakistani electronic media and social media began to analyze the possible causes for the accident.

The Indian Defence Ministry's statement of 11 March acknowledging the accidental launch of the missile was immediately picked up by Pakistani media which sought to undermine India's claims of being a responsible nuclear state which had resulted in India-specific exemptions such as the 2008 US-India nuclear deal, denied to Pakistan. A 11 March article in *Daily Pakistan*, sarcastically titled as "A responsible nuclear state?" declared with some satisfaction that the Indian Defence Ministry had "admitted firing a missile into Pakistan" in what essentially amounted to an "Indian confession," a day after the Pakistani DG-ISPR's press conference.[57]

Pakistan's thriving community of social media channels also took notice of, and attempted to glean the Indian response to, the DG-ISPR's press conference. Pakistani journalist Arzoo Kazmi questioned Indian columnist and political commentator Sumit Peer on India's response to the DG-ISPR statement on her private YouTube channel on 11 March with Peer responding that clarity would only emerge once India made an official statement.[58]

Pakistani journalist Makhdoom Shahab-ud-Din also weighed in on the missile accident on his private YouTube channel on 11 March and highlighted the Indian Defence Ministry's statement admitting its error as an Indian admission of its incompetent missile systems. He also jubilantly informed his audience that the Pakistan Air Force, through its monitoring of the errant missile, had proved its superiority in the entire region and that Indian missile platforms had been globally discredited as a result of the accident. Shahab-ud-Din also lampooned India's broader safety culture by declaring Indian fighter aircraft as "flying coffins" and reminding his viewers of the accidental Indian shooting down of its own helicopter during the Balakot airstrikes. He ended with a warning that this lax safety culture was now a danger for the entire region and had put paid to Indian claims of being a regional counterweight to China.[59]

In another video a day later, Shahab-ud-Din declared that the crashing of the BrahMos missile into Pakistan was the greatest opportunity for Pakistan and China as Pakistan now had access to the highly sophisticated chips and processors powering the missile and could reverse engineer it. As an example, he discussed former Pakistani Prime Minister Nawaz Sharif's disclosure regarding the Pakistani reverse engineering of an errant American Tomahawk cruise missile that had strayed from its intended target in Afghanistan into Pakistan in 1998 which contributed towards production of Pakistan's Babur cruise missile.[60]

Indian Media's Praise for Pakistan's Restraint, Call for Joint Probe, Deliberate Firing

Pakistani journalist Ribaha Imran posted an online video of her interview with eight civilians in which a majority expressed satisfaction over India's admission of the accidental launch and the maturity displayed by both countries in dealing with the crisis. Some criticism was also directed at the Pakistani military for its perceived 24-hour delay in announcing the missile's violation of Pakistani airspace.[61]

Other articles also focused on the criticism leveled by Indian journalists on the Indian government for its "missile fiasco"[62] and on their praise for Pakistan for its "wise decision" of acting "prudently in this matter."[63]

> A journalist from *Indian Express*, Nirupama Subramanian said, "Wish India had acknowledged this missile fiasco immediately after it happened, instead of keeping quiet for 48 hours after the incident, and making a statement only after Pakistan announced it"...Shekhar Gupta, *ThePrint* Founder, wrote on Twitter, "Think for a moment if a Pakistani cruise missile had strayed into India instead. Govt might've kept its counsel but our warrior TV channels would've gone ballistic—with warheads, of course." Praveen Swami from *The Print* said, "For anyone underestimating the seriousness of one nuclear-weapons state accidentally firing cruise missile at another nuclear-weapons state: consider, slowly and carefully, the three words Mutually Assured Destruction....The reason this was handled calmly was we're not in a crisis, and no-one has itchy trigger fingers."[64]

The *Daily Islamabad Post* also gave prominent coverage to praise for Pakistan's restraint by Indian journalists including Sushant Singh, former Deputy Editor of the *Indian Express*.

> British Broadcasting Corporation (BBC) in an analysis posted on its website said, "Any object entering the airspace of any country is considered

> an attack. In such a case, the defense rules and regulations say that in case of a missile, you should also retaliate. So not doing so on Pakistan's part is a wise decision as well." These are the words of Sushant Singh, a senior fellow at the Center for Policy Research in India, who praised Pakistan's "resolved response" to the accidental landing of an Indian BrahMos missile on March 9, 2022 in Mian Channu, Pakistan....many Indian journalists and military experts are giving credit to the Pakistani government, especially at a time when it took two days for India to admit....What would have happened, Sushant asked, if the missile had hit a big city; thankfully, no casualty took place. He said that Pakistan's defense system is much better. "Even so, no defense system can determine whether a cross-border missile has been fired accidentally." Praveen Sohne, editor of the Indian military magazine Force, told the BBC: Pakistan has "acted prudently in this matter which should take place in a democratic government".[65]

Pakistan Today editorialized on 13 March urging the Pakistan government to continue "pursuing the path of diplomacy." It also made the case for a joint investigation, even as it was skeptical as to whether India would allow it. The paper also couldn't hide the lurking suspicion shared by sections of the Pakistani policy elite and its media community that the supposedly inadvertent firing could have also been a deliberate Indian ploy to test Pakistani defenses, a sentiment that had been buttressed by the claimed thwarting by the Pakistan Navy of an Indian submarine's attempt to enter Pakistani waters on 1 March.[66]

> Pakistan had no means of knowing that the missile was not carrying a nuclear payload until it landed near Mian Channu. It is also fortunate that it landed in an empty area, and did not land in a built-up area. It is also fortunate that its path did not cross that of any civilian airliner. These pieces of good fortune cannot be ascribed to any wisdom or prudence, but were the result of mere chance...A joint investigation would also lead to a means of developing procedures for the two countries to follow in the event of such accidents.....It would not be beyond India to fire off an expensive blank as a test.[67]

The same newspaper in another editorial two days later, reiterated the Pakistani call for a joint probe and talks on other matters as the "the logical way forward." The editorial board also excoriated the US for its India-friendly statement on the missile accident while praising China.

> Following India's missile misadventure that could've ended in disaster, possibly creating a serious diplomatic crisis, the USA has weighed in

> on the matter and its response, expectedly, is tone deaf and needlessly heavily in favor of Delhi. In the absence of a detailed inquiry, the USA stating unequivocally that this cannot be anything but an accident is a biased statement stemming from its less than cordial current relations with Islamabad. Alternatively, China has taken a much more nuanced stance, encouraging both countries to hold talks and a joint probe into the "accidental" missile firing should be held. It is quite understandable why Pakistan would like to join the enquiry into the straying of an Indian BrahMos missile into Pakistan. For one, Pakistan had no means of knowing that the missile was not carrying a nuclear payload until it landed near Mian Channu. Such incidents must be avoided, because the other side might respond aggressively.[68]

Attiya Munawer, opined on 17 March in *Pakistan Today* that India needed to allow the Pakistani request for a joint investigation.

> Be it India's so-called surgical strike or Indian submarines, or encroachments on Pakistani waters, Pakistan has always acted with tolerance, but it seems that the irresponsible behaviour of the Indian Army will jeopardize not only the region but also world peace....The world should take serious note of this and demand an investigation, although the Indian Defence Minister has admitted his mistake and announced an investigation into the incident. However, this should be done through a joint operation with Pakistan. So that it is clear that this was a human error or a deliberate conspiracy because even before this, India has on several occasions fanned the flames of war and provocations....However, if Pakistan had not shown extraordinary restraint on this serious border violation and had reacted aggressively using its inalienable right to defend its international borders and airspace, war could have erupted between the two nuclear powers.[69]

Initial Results of India's Court of Inquiry: Pakistan's Retaliation Planned with Armed Warhead

The Pakistani media also reported on the initial results of the Indian CoI, including its announcement of the culpability of a Group Captain. In another video on 23 March, journalist Makhdoom Shahab-ud-DIn praised the Pakistani ISPR for the measured tone of its press conference which had cornered India into accepting the accidental launch and initiating a high-level inquiry. He also informed viewers about the article in *Bloomberg* regarding the Pakistani preparation for a retaliation and ventured (without providing evidence) that the retaliation was planned with a missile with an armed warhead. Shahab-ud-Din also made much of the statement on the incident released by the Organization of Islamic Countries (OIC) which criticized India. He also worried about the culpability of the Group Captain raising questions about whether his malevolence or incompetence was responsible for the accident and whether he was being scapegoated for a grave lapse which was more of a systemic problem rather than an individual's fault with the former being left unaddressed.[70]

Haroon Hayder, a senior news writer for *ProPakistani* (an independent news publisher) relayed reports in the Indian media in his article on 25 March that four IAF personnel were under investigation for the accidental launch.

> Four top-ranked officials of the Indian Air Force (IAF) are being investigated for their involvement in the accidental misfire of the BrahMos missile into Pakistan earlier this month...Reports have confirmed that one of the officers under probe is a Group Captain. Note here that Group Captain or Commanding Officer is the sixth-highest rank in the IAF... The initial investigation of COI had found that it was a human error and not a technical fault that led to the missile misfire.[71]

The Pakistani media also gave prominent coverage to the Philippines questioning India over its missile technology, including this 7 April article in *The News:*

> The Philippines has asked New Delhi for a clarification on the BrahMos missile it is buying from India days after the defence ministry admitted to a missile having been accidentally fired into Pakistan...Indian ambassador to the Philippines, Shambhu S. Kumaran, said this in the course of a virtual talk on the partnership between the two countries organised by the Ananta Centre on April 5...."Yes, I did have an opportunity to interact with [Philippines] defence secretary [Delfin Negrillo] Lorenzana and I clarified and our Ministry of Defence clarified... that there was no technical issue as far as we could understand it....Lorenzana was the official who signed the $375 million (Rs69.5 billion) deal for the Philippine Marines, with India's BrahMos Aerospace, for the former to acquire three batteries of the BrahMos cruise missile....The Philippines is the first country to buy the BrahMos missile.[72]

Court of Inquiry's Final Verdict, Cover-up to Hide Command-and-Control Deficiencies

In late August 2022, the Pakistani media reported on the Indian Court of Inquiry's key finding holding a Group Captain and two Wing Commanders responsible for deviation from Standard Operating Procedures (SOPs), the Indian government's subsequent decision to fire them, Pakistan's rejection of the inquiry's findings and its reiteration of its demand for a joint probe.[73] *ARY News* channeled the Pakistani establishment's dissatisfaction with the Indian probe and its suspicion that India was covering up systemic deficiencies in its command and control by blaming the accident on human error.

> Pakistan on Wednesday categorically rejected India's "purported closure" of the highly irresponsible incident of the firing of a rogue supersonic missile into Pakistani territory and reiterated its demand for a joint probe.... The spokesperson said India had not only failed to respond to Pakistan's demand for a joint inquiry, but has also evaded the questions raised by Pakistan regarding the command and control system in place in India, the safety and security protocols and the reason for India's delayed admission of the missile launch. The spokesperson further said that systemic loopholes and technical lapses of serious nature in the handling of strategic weapons could not be covered up beneath the veneer of individual human error.[74]

A 26 August editorial in *The Dawn* also channeled the Pakistani dissatisfaction with the findings of the Indian Court of Inquiry and reiterated the demand for a joint probe.

> The intrusion of an Indian nuclear-capable BrahMos missile into Pakistani territory in March was a grave incident that could have had disastrous consequences. Fortunately, Pakistan handled the incident in a mature fashion and it was a relief that the projectile caused no loss of life when it crashed in Mian Channu, Punjab. Wrapping up the investigation into the episode, the Indian Air Force recently sacked three officers for the "violation of Standard Operating Procedures" that caused the "accidental firing of the missile". However, Pakistan has rejected these findings, with the Foreign Office terming the probe "unsatisfactory, deficient and inadequate". Indeed, many questions remain unanswered about the chilling event....As per the PAF (Pakistan Air Force), at the time the missile was picked up there were several airliners in the surrounding airspace. The FO (Foreign Office), therefore, is right in demanding a joint probe.[75]

And finally, the Pakistani press also criticized the Indian government and the Indian media for purportedly mis-characterizing the International Atomic Energy Agency (IAEA) Director General's comment that the BrahMos missile misfire was not an issue of particular concern to the IAEA.[76] Interestingly, an editorial in the *Pakistan Observer* on 18 November claimed that the errant BrahMos missile had landed near a nuclear facility in Pakistan.

> On Monday, some sections of Indian media quoted international nuclear watchdog International Atomic Energy Agency (IAEA) as saying that it did not see the incident (missile accident) as any cause of "specific concern"....Firing a super-sonic missile into Pakistani territory and, above all, its landing close to a nuclear facility was an unforgivable crime but Pakistan, as per its well thought-out policy, went for a cool-minded approach as it always tried to de-escalate tension between the two countries....Mere suspension of officials is not enough and India must satisfy Pakistan on points of concern like the underlying intentions, technical features and reliability of the missile system, safety, security and nuclear command and control protocols and the presence of rogue elements in the Indian military.[77]

The Indian Media—Critical but Supportive

An examination of the media discourse in both India and Pakistan over the missile accident reveals the emergence of two different narratives, both with their own internal heterogeneity. In the immediate aftermath of the accident, the Indian media was faced with the challenge of piecing together the events that led to the misfiring in an environment of severely limited information from the government. The resulting strategy of a major section of the Indian media was a combination of careful parsing of official statements, relying on anonymous leaks both from serving and retired government officials and informed speculation by veteran journalists with decades of experience. Thus, the media was quick to notice and highlight the shift in government's claim from technical error in initial statement on 11 March to possible human error in Rajnath Singh's statement on 15 March.

However, the media's attempt to uncover and publicize information pertaining to the accident was also constrained by a powerful nationalistic imperative to prevent the accident from becoming a major international embarrassment to the Indian government or to provide an opening for Pakistan to dent India's image as a responsible state. So even as the media criticized the Indian government's initial delay in issuing a formal statement on the accident and called for a swift and impartial investigation to affix responsibility, it was also careful to repeatedly specify that such an investigation be conducted solely by India with no role for Pakistan. Even as sections of the media praised Pakistan for displaying maturity by not retaliating to the errant Indian missile, they also attempted to moderate the seriousness of the accident by claiming that the BrahMos missile was not nuclear-capable.

The supportive tone of the Indian media's reporting on the findings of the Indian Court of Inquiry and its general satisfaction with punitive action against three culpable officers without more forceful and continuous calls for systemic reform of command and control systems also reinforces its critical (but ultimately supportive) stance for the government's actions pertaining to the accident. Only a minority faction in the Indian media adopted a jubilant tone over the missile's breach of Pakistani air defense systems, claiming the launch to be a deliberate act to test Pakistan's response spectrum.

The Pakistani Media—Vindicated, Suspicious and Dissatisfied

The Pakistani media's coverage of the missile accident was shaped by coverage of the accident by the Indian media, nationalism, trust deficit with India, Indo-skepticism and Indo-pessimism. Even as the media severely excoriated the Indian government for its delay in issuing a satisfactory explanatory statement in the immediate aftermath of the accident and even as it expressed intense dissatisfaction over the generic brevity of those statements, it also took note of the Indian media's praise for Pakistan's decision to not retaliate with its own missile. For the Pakistani media (which gave prominent coverage to comments by Indian journalists praising Pakistan's restraint), such praise was vindication of Pakistan's self-perception as a responsible state in the region. This self-perception was also bolstered by coverage in the international press (*Bloomberg*) regarding Pakistan's decision to consider but not carry out its plans for retaliation with a similar missile.

The Indian government's "confession" of responsibility for the accident was also a validation of Pakistan's allegations that there was massive dysfunction behind the thin veneer of technological sophistication and organizational competence that India claimed for its command and control systems. Adopting a nationalistic stance similar to that of the Indian media, the Pakistani media threw its support behind the Pakistani government's demand for a detailed explanation and for a joint probe by the Indian government. It also amplified the sentiments of the Pakistani foreign policy elite of India as an irresponsible state in the region and deserving of international pressure and closer scrutiny of its command and control systems. In doing so, it de-privileged a more forceful and concerted questioning of the Pakistani government for not being able to shoot down the missile while it was in Pakistani airspace, a line of critique which would have questioned the competence of Pakistani air defense systems and personnel operating them. The Pakistani media also focused repeatedly on the "nuclear-capable" feature of the BrahMos missile in its bid to elevate the seriousness of the missile misfire and bolster its demand for a joint probe. Co-existing uneasily beside the Pakistani allegation of massive Indian technical and organizational incompetence was the deep distrust of the current Indian government and memories of its previous punitive actions, especially the aerial strike at Balakot. This distrust led to analyses that India's firing of the missile was not just an accident resulting from a technical or human error as claimed at various times by the Indian government but could have been a deliberate attempt by an extremist government to test Pakistan's operational capabilities. The distrust also led to deep dissatisfaction in the Pakistani media over the final conclusion of India's Court of Inquiry which was viewed as a massive cover-up of the true (and systemic) extent of India's failings by scapegoating the three IAF officers. A minority faction in the Pakistani media also attempted to reframe the Indian ingress (whether accidental or deliberate) of Pakistani airspace as an ideal opportunity to reverse-engineer it and to share its technological contents with China as a way of punishing India.

Ultimately, the coverage of the missile accident by the Indian and Pakistani media aligned closely with the minimalist and maximalist interpretations of the incident by their respective governments. The minimalist interpretation advanced by the Indian state (and supported for most part by the Indian media) was of an isolated accidental release of an unarmed, conventional cruise missile in an otherwise foolproof command and control system caused due to the individual errors of judgment by three IAF officers. This framing of the accident resulted in a solution that was necessarily limited (Court of Inquiry) and thus in India's total control. This Indian narrative failed to agree with a rival Pakistani maximalist narrative of the missile accident as the result of a massive dysfunction in the error-prone and competence-challenged, command and control systems of an irresponsible state which could not be trusted with the safe handling of a nuclear-capable missile with tremendous destructive power. This Pakistani framing of the accident opened the space for more radical solutions (joint probe) and for the internationalization of the accident.

Notes

1 Sami Khan, "Aircraft Crashes in PAK's Khanewal; Air Force Jet Suspected, No Casualties," *International Business Times,* March 9, 2022, accessed January 3, 2023, https://www.ibtimes.co.in/aircraft-crashes-paks-khanewal-air-force-jet-suspected-no-casualties-details-846433
TV9, "Pakistan Plane Crash:....," *TV9 Bharatvarsh,* March 9, 2022, accessed January 3, 2023, https://www.tv9hindi.com/world/pakistan-news/pakistan-plane-crash-news-pakistan-air-force-army-plane-crash-in-punjab-province-mian-channu-1104836.html

2 F P Explainers, "India Accidentally Fires Missile into Pakistan: What and How Did It Happen?," *Firstpost,* March 12, 2022, accessed January 4, 2023, https://www.firstpost.com/india/india-accidentally-fires-missile-into-pakistan-what-and-how-did-it-happen-10452681.html

3 "Unguided Missile: The Hindu Editorial on a Missile Malfunction and Pakistan's Joint Probe Demand," *The Hindu,* March 13, 2022, accessed January 4, 2023, https://www.thehindu.com/opinion/editorial/unguided-missile-the-hindu-editorial-on-a-missile-malfunction-and-pakistans-joint-probe-demand/article65221734.ece

4 Seema Guha, "India's Accidental Firing of Unarmed Missile into Pakistan Raises Questions of Safety Protocols," *Outlook,* March 13, 2022, accessed January 5, 2023, https://www.outlookindia.com/national/india-accidental-firing-of-missile

-into-pakistan-questions-of-safety-protocols-news-186688?utm_source=related_story
5 Dinakar Peri, "Explained | An Accidental Missile Launch," *The Hindu*, March 16, 2022, sec. India, accessed January 5, 2023, https://www.thehindu.com/news/national/an-accidental-missile-launch/article65228497.ece
6 Dinakar Peri, "Explained | An Accidental Missile Launch," *The Hindu*, March 16, 2022, sec. India, accessed January 5, 2023, https://www.thehindu.com/news/national/an-accidental-missile-launch/article65228497.ece
7 "Unguided Missile: The Hindu Editorial on a Missile Malfunction and Pakistan's Joint Probe Demand," *The Hindu*, March 13, 2022, accessed January 4, 2023, https://www.thehindu.com/opinion/editorial/unguided-missile-the-hindu-editorial-on-a-missile-malfunction-and-pakistans-joint-probe-demand/article65221734.ece
8 T. O. I. Edit, "Truly Scary Moment: IAF Must Thoroughly Investigate Accidental Missile Launch. India, Pak Must Develop Risk Reduction Protocols," *The Times of India*, March 13, 2022, accessed January 5, 2023, https://timesofindia.indiatimes.com/blogs/toi-editorials/truly-scary-moment-iaf-must-thoroughly-investigate-accidental-missile-launch-india-pak-must-develop-risk-reduction-protocols/
9 "India, Pakistan Have Acted with Maturity in Wake of Missile Incident. Delhi Must Conduct Thorough Probe, Fix Responsibility," *The Indian Express,* March 15, 2022, accessed January 5, 2023, https://indianexpress.com/article/opinion/editorials/unmaking-a-crisis-india-pakistan-missile-7819887/
10 Rahul Bedi, "India Needs to Come Clean on 'Accidental Firing of Missile' to Avert Conflict with Pakistan," *The Wire,* March 12, 2022, accessed January 5, 2023, https://thewire.in/security/india-needs-to-come-clean-on-accidental-firing-of-missile-to-avert-conflict-with-pakistan
11 Rahul Bedi, "After Missile Fiasco, Indian Officials Hope Spin Will Deflect Awkward Questions," *The Wire,* March 15, 2022, accessed January 5, 2023, https://thewire.in/government/missile-fiasco-india-pakistan-awkward-questions-spin
12 CPR, "Sushant Singh," *Centre for Policy Research*, accessed January 7, 2023, https://cprindia.org/people/sushant-singh/
13 Sushant Singh, "A 'Broken Arrow' Raises Serious Questions," *Deccan Herald,* March 13, 2022, accessed January 7, 2023, https://www.deccanherald.com/india/a-broken-arrow-raises-serious-questions-1090769.html
14 Abhinandan Mishra, "Pak Fails to Intercept India's Missile, Unease in GHQ," *The Sunday Guardian Live*, March 12, 2022, accessed January 7, 2023, https://sundayguardianlive.com/news/pak-fails-intercept-indias-missile-unease-ghq
15 Arzoo Kazmi, "Pak Army's Restrained Response to the Indian Missile Incident Averted Grave Consequences," *News9live,* March 13, 2022, accessed January 8, 2023, http://www.news9live.com/india/pak-armys-restrained-response-to-the-indian-missile-incident-averted-grave-consequences-158830
16 "BrahMos Can Soon Hit Deep inside China; Hypersonic Missile Strike Range to Be Extended to 800 Km," *Hindustan Times,* March 14, 2022, accessed January 8, 2023, https://www.youtube.com/watch?v=u7cJyrLfPA4
17 Bureau, "China Meddles in Indian Affairs, Urges Pakistan to Seek.....," *Republic World,* March 14, 2022, accessed January 8, 2023, https://www.republicworld.com/world-news/pakistan-news/china-meddles-in-indian-affairs-urges-pakistan-to-seek-in-depth-probe-in-missile-episode-articleshow.html/
18 Staff, "'India Planning to Invade, Yogi Celebrating Victory': Pakistani Experts Make Wild Claims about Alleged 'Indian Missile,'" *OpIndia*, March 11, 2022, accessed January 9, 2023, https://www.opindia.com/2022/03/india-preparing

-for-invasion-yogi-adityanath-celebrating-election-victory-pakistani-experts -wild-theories-about-alleged-indian-missile/

19 OpIndia Staff, "Netizens Share Hilarious Posts after India Admits to Accidentally Firing a Missile into Pakistan," *OpIndia*, March 11, 2022, accessed January 9, 2023, https://www.opindia.com/2022/03/netizens-have-ameme-fest-after-india -admits-firing-missile-into-pakistan/

20 "Top Air Force Officer 'Blameworthy' in Accidental Missile Firing on Pak: Report," *Hindustan Times*, March 23, 2022, Accessed January 11, 2023, https://www.hindustantimes.com/india-news/accidental-missile-firing-incident -prima-facie-fingers-pointed-at-group-captain-101648015107516.html

21 Praveen Swami, "Missiles, Malfunctions, Nuke Myths: Lessons for India, Pakistan from BrahMos Mistake," *ThePrint*, March 16, 2022, accessed January 11, 2023, https://theprint.in/opinion/security-code/missiles-malfunctions-nuke -myths-lessons-for-india-pakistan-from-brahmos-mistake/875149/
Missiles, "Malfunctions, Nuke Myths: Lessons for India, Pakistan from BrahMos mistake-----Praveen Swami," *The Print (video*), March 17, 2022, https://www.youtube.com/watch?v=SgUVysE-psc
Missiles, "Malfunctions, Nuke Myths: Lessons for India, Pakistan from BrahMos. Mistake," www.youtube.com *The Print*, March 17, 2022, accessed January 11, 2023, https://www.youtube.com/watch?v=SgUVysE-psc

22 PTI, "On Indian Missile Landing In Pak, US Supports Delhi's Stand," *NDTV. Com*, March 15, 2022, accessed January 11, 2023, https://www.ndtv.com/india -news/what-us-said-on-indias-accidental-missile-firing-into-pakistan-2823163
Agencies, "'We Have No Indication..': What US Said on India's 'accidental' Missile Firing," *Mint*, March 15, 2022, accessed January 11, 2023, https://www .livemint.com/news/world/we-have-no-indication-what-us-said-on-india-s-accidental-missile-firing-into-pakistan-11647310974716.html

23 Sen Ranjan Sudhi and Faseeh Mangi, "Errant Indian Missile Nearly Led to Pakistan Retaliatory Strike," *Bloomberg.Com*, March 15, 2022, accessed January 12, 2023, https://www.bloomberg.com/news/articles/2022-03-15/ errant-indian-missile-nearly-led-to-pakistan-retaliatory-strike

24 Som Vishnu, "Indian Missile Accident Nearly Led to Pak Retaliation: Report," *NDTV.Com*, March 16, 2022, accessed January 12, 2023, https://www.ndtv .com/india-news/indian-missile-accident-nearly-led-to-pak-retaliation-report -2825860
Agencies, "Pak Nearly Launched Retaliatory Strike after India's Accidental Missile Launch," *Mint*, March 16, 2022, accessed January 15, 2023, https:// www.livemint.com/news/world/pakistan-nearly-launched-retaliatory-strike-in -response-to-india-s-accidental-missile-launch-11647424134509.html
Scroll Staff, "Pakistan Nearly Launched Retaliatory Strike after India's Missile Misfire: Report," *Scroll.In*, March 15, 2022, accessed January 15, 2023, https:// scroll.in/latest/1019598/pakistan-nearly-launched-retaliatory-strike-after -indias-missile-misfire-report
Web Desk, "Pakistan Planned Retaliatory Strike after India's Missile Fiasco: Report," *The Week*, March 16, 2022, accessed January 15, 2023, https://www .theweek.in/news/world/2022/03/16/pakistan-planned-retaliatory-strike-after -indias-missile-fiasco-report.html

25 "How Did India Accidentally Fire a Missile Into Pakistan? | NDTV-No Spin," March 14, 2022, accessed January 15, 2023, https://www.youtube.com/watch?v =DqCT8QONTVw

26 "How Did India Accidentally Fire a Missile Into Pakistan? | NDTV-No Spin," March 14, 2022, accessed January 15, 2023, https://www.youtube.com/watch?v =DqCT8QONTVw

27 "Indian BrahMos with 'a Mind of Its Own' Lands in Pakistan & CBMs to Prevent Unintended Escalation," *The Print-Cut the Clutter*, March 16, 2022, accessed January 15, 2023, https://www.youtube.com/watch?v=c7aJrZdKZm8
28 "Indian BrahMos with 'a Mind of Its Own' Lands in Pakistan & CBMs to Prevent Unintended Escalation," *The Print-Cut the Clutter*, March 16, 2022, accessed January 15, 2023, https://www.youtube.com/watch?v=c7aJrZdKZm
29 Krishn Kaushik, "Explained: When a Missile Misfires," *The Indian Express*, March 12, 2022, accessed January 16, 2023, https://indianexpress.com/article/explained/explained-when-a-missile-misfires-india-pakistan-brahmos-7815962/
30 Rahul Bedi, "Did Rajnath Singh Subtly Suggest 'Human Error' Led to Accidental Missile Firing?" *The Wire*, March 17, 2022, accessed January 16, 2023, https://thewire.in/security/india-accidental-missile-firing-human-error-rajnath-singh
31 ANI, "Pakistan's 'missile' Fails to Reach Target, Crashes in Sindh," *ThePrint*, March 18, 2022, accessed January 16, 2023, https://theprint.in/world/pakistans-missile-fails-to-reach-target-crashes-in-sindh/878464/
ANI, "'Missile Test' by Pakistan Fails," *ThePrint*, March 17, 2022, accessed January 17, 2023, https://theprint.in/world/missile-test-by-pakistan-fails/878270/
News Desk, "Missile Test by Pakistan Fails, Crashes in Sindh," *IndiaTV*, March 18, 2022, accessed January 17, 2023, https://www.indiatvnews.com/news/world/pakistan-missile-test-fails-crashes-in-sindh-2022-03-18-764899
Staff, "Pakistan: Unidentified Object Falls from Sky, a Failed Missile Test Speculated," *OpIndia*, March 17, 2022, accessed January 17, 2023, https://www.opindia.com/2022/03/pakistan-unidentified-object-falls-from-sky-a-failed-missile-test-speculated/
32 Syed Ali Mujtaba, "Demystifying India's Missile Mis-Launch," *The Siasat Daily*, March 19, 2022, accessed January 18, 2023, https://www.siasat.com/demystifying-indias-missile-mis-launch-2293287/
33 "Top Air Force Officer 'Blameworthy' in Accidental Missile Firing on Pak: Report," *Hindustan Times*, March 23, 2022, accessed January 11, 2023, https://www.hindustantimes.com/india-news/accidental-missile-firing-incident-prima-facie-fingers-pointed-at-group-captain-101648015107516.html
TNN, "Lens on Group Captain for Missile Misfire," *The Times of India*, March 24, 2022, accessed January 18 2023, https://timesofindia.indiatimes.com/india/lens-on-group-captain-for-missile-misfire/articleshow/90406580.cms?utm_source=contentofinterest&utm_medium=text&utm_campaign=cppst
34 ANI, "IAF Officer's Probe Suggests Prima Facie Group Captain Involved in Accidental BrahMos Firing," *ThePrint*, March 23, 2022, accessed January 18, 2023, https://theprint.in/india/iaf-officer-probing-accidental-missile-firing-incident-prima-facie-fingers-pointed-at-group-captain/884671/
35 "Brahmos Missile Probe | Pakistan's Lies Busted As Investigation Reveals 'Human Error' | Latest News," *TimesNow*, March 24, 2022, accessed January 18, 2023, https://www.timesnownews.com/videos/times-now/india/brahmos-missile-probe-pakistans-lies-busted-as-investigation-reveals-human-error-latest-news-video-90411848
36 TN National Desk, "Accidental BrahMos Firing Incident: Inquiry Holds More than Two Officers Responsible," *TimesNow*, April 11, 2022, accessed January 19, 2023, https://www.timesnownews.com/india/accidental-brahmos-firing-incident-inquiry-holds-more-than-two-officers-responsible-article-90768987
37 Rajat Pandit, "BrahMos Unit Chief, Others May Face Action over 'Misfire.'" *The Times of India*, April 11, 2022, accessed January 19, 2023, https://timesofindia.indiatimes.com/india/brahmos-unit-chief-others-may-face-action-over-misfire/

articleshow/90767344.cms?utm_source=contentofinterest&utm_medium=text&utm_campaign=cppst

38 Abhishek Srivastava, "Did India's Missile Misfire Expose the Banana Military of the Banana Republic - Pakistan?" *NewsGram,* April 27, 2022, accessed January 19, 2023, https://www.newsgram.com/defense/2022/04/27/did-indian-missile-misfire-expose-the-banana-military-of-the-banana-republic-pakistan

39 "India's BrahMos Missile Accident Exposes Pakistan's Weakness? China-Made Weapons Can't Stop BrahMos? Editorji," August 29, 2022, accessed January 19, 2023, https://www.youtube.com/watch?v=qu8I6CIvS4U

40 Editorji News Desk, "Indian Missile Accident Exposes Pakistan's Weakness? Chinese Weapons Can't Stop BrahMos? | Doom Tech," *Editorji,* August 29, 2022, accessed January 20, 2023, https://www.editorji.com/in-depth/the-hook/indian-missile-accident-exposes-pakistan-s-weakness-chinese-weapons-can-t-stop-brahmos-doom-tech-1661769646041

41 "India's BrahMos Missile Accident Exposes Pakistan's Weakness? China-Made Weapons Can't Stop BrahMos? Editorji," August 29, 2022, accessed January 19, 2023, https://www.youtube.com/watch?v=qu8I6CIvS4U

42 "India Sacks 3 Officers for Firing Accidental Missile to Pakistan," August 25, 2022, accessed January 19, 2023, https://www.youtube.com/watch?v=BU2gRQnsGJY

43 Agencies, "Three IAF Officers Sacked for BrahMos Missile Misfire over Pakistan," *The Times of India,* August 23, 2022, accessed January 20, 2023, https://timesofindia.indiatimes.com/india/three-iaf-officers-sacked-for-missile-misfire/articleshow/93735663.cms

44 Tribune News Service, "IAF Sacks 3 Officers for Accidental Firing of Brahmos Missile That Landed in Pakistan," *The Tribune,* August 23, 2022, https://www.tribuneindia.com/news/nation/iaf-sacks-3-officers-for-accidental-missile-fire-424654

The Wire Staff, "Three IAF Officers Sacked for 'accidental Firing' of Brahmos Missile That Landed in Pak," *The Wire,* August 23, 2022, accessed January 2023, https://thewire.in/security/three-iaf-officers-sacked-accidental-firing-brahmos-pakistan

Rahul Singh, "IAF Sacks Three Officers for BrahMos Misfire into Pakistan in March," *Hindustan Times,* August 23, 2022, accessed January 20, 2023, https://www.hindustantimes.com/india-news/iaf-sacks-three-officers-for-brahmos-misfire-into-pakistan-in-march-101661263805711.html

45 Kalyan Ray, "Three IAF Officers Sacked for Accidental Firing of Brahmos Missile That Landed in Pakistan," *Deccan Herald,* August 23, 2022, accessed January 20, 2023, https://www.deccanherald.com/india/three-iaf-officers-sacked-for-accidental-firing-of-brahmos-missile-that-landed-in-pakistan-1138593.html

46 DHNS, "Action for Firing BrahMos Missile into Pak Welcome," *Deccan Herald,* September 3, 2022, accessed January 23, 2023, https://www.deccanherald.com/opinion/editorial/action-for-firing-brahmos-missile-into-pak-welcome-1141656.html

47 "Pakistan Demands Explanation after Indian Projectile Crashes Near Mian Channu," *SAMAA,* March 10, 2022, accessed January 23, 2023. Pakistan demands explanation after Indian projectile crashes near Mian Channu - Archives - SAMAA (samaaenglish.tv)

"DGISPR Demands Explanation after Indian Projectile Crashes near Mian Channu," *ARY News,* March 11, 2022, accessed January 23, 2023, https://www.youtube.com/watch?v=qhy9GsZl0Tk

Web Desk, "Indian Projectile Entered Pakistani Airspace, Caused Damage: DG ISPR," *The News,* March 10, 2022, accessed January 23, 2023, https://www

.thenews.com.pk/latest/940308-indian-projectile-entered-pakistani-airspace-caused-damage-dg-ispr

48 Dawn.com, "India 'Deeply Regrets Accidental' Firing of Missile inside Pakistan," *DAWN,* March 11, 2022, accessed January 24, 2023, https://www.dawn.com/news/1679443

49 Dawn.com, "India 'Deeply Regrets Accidental' Firing of Missile inside Pakistan," *DAWN,* March 11, 2022, accessed January 24, 2023, https://www.dawn.com/news/1679443

50 Baqir Sajjad Syed, "World Asked to Verify Indian Claims about Missile," *DAWN,* March 12, 2022, accessed January 24, 2023, https://www.dawn.com/news/1679534

51 "India Admits Mistake of Firing Missile at Pakistan in Mian Channu," *Dunya News,* March 11 2022, accessed January 24, 2023, https://www.youtube.com/watch?v=SOd4MEkrCAg

52 Naveed Siddiqui, "Why Did India Fail to Immediately Inform about Accidental Missile Launch? Asks Pakistan," *DAWN.COM*, March 12, 2022, accessed January 24, 2023, https://www.dawn.com/news/1679603

53 Naveed Siddiqui, "Why Did India Fail to Immediately Inform about Accidental Missile Launch? Asks Pakistan," *DAWN.COM*, March 12, 2022, accessed January 24, 2023, https://www.dawn.com/news/1679603
Anwar Iqbal Syed | Baqir Sajjad, "Pakistan Demands Joint Probe into Indian Missile Fiasco," *DAWN.COM*, March 13, 2022, accessed January 24, 2023, https://www.dawn.com/news/1679725
Kamran Yousaf, "Pakistan Seeks Joint Probe into Indian Missile Fiasco," *The Express Tribune,* March 12, 2022, accessed January 24, 2023, https://tribune.com.pk/story/2347607/pakistan-seeks-joint-probe-into-indian-missile-fiasco

54 Baqir Sajjad Syed, "World Asked to Verify Indian Claims about Missile," *DAWN,* March 12, 2022, accessed January 24, 2023, https://www.dawn.com/news/1679534

55 Naveed Siddiqui, "Why Did India Fail to Immediately Inform about Accidental Missile Launch? Asks Pakistan," *DAWN.COM*, March 12, 2022, accessed January 24, 2023, https://www.dawn.com/news/1679603

56 Anwar Iqbal Syed | Baqir Sajjad, "Pakistan Demands Joint Probe into Indian Missile Fiasco," *DAWN.COM*, March 13, 2022, accessed January 24, 2023, https://www.dawn.com/news/1679725

57 Web Desk, "A Responsible Nuclear State? India Says 'Accidentally' Fired a Supersonic Missile into Pakistan," *Daily Pakistan,* March 11, 2022, accessed January 24, 2023, https://en.dailypakistan.com.pk/11-Mar-2022/deeply-regret-table-india-admits-firing-missile-into-pakistan

58 Arzoo Kazmi, *Pakistan Hit by "Unarmed Supersonic Missile" Allegedly Fired by India,* March 11, 2022, accessed January 24, 2023, https://www.youtube.com/watch?v=rfURgzABAvg

59 Makhdoom Shahab-ud-Din, "Pakistan Air Force At His Best |Indian Defense Ministry Backs DG ISPR On Brahmos Supersonic Missile," March 11, 2022, accessed January 24, 2023, https://www.youtube.com/watch?v=ucr_KQ52B6U

60 Makhdoom Shahab-ud-Din, "An Opportunity For Pakistan from Brahmos Missile Intercepted by Pakistan Air Force," March 12, 2022, accessed January 25, 2023, https://www.youtube.com/watch?v=L01kwXAlWVc

61 Ribaha Imran, "Brahmos Strike On Pakistan Was a Mistake Says India - Pakistani Reaction," March 13, 2022, accessed January 26, 2023, https://www.youtube.com/watch?v=_mdHk46yNeM

62 Web Desk, "Indian Journalists Criticise Missile Fiasco by Nuclear-Armed India," *News 360,* March 12, 2022, accessed January 25, 2023, https://news360.tv/en/latest/indian-journalists-criticise-missile-fiasco-by-a-nuclear-armed-india/

63 Web Desk, "Indian Journalists, Military Experts Hailing Pakistan for Prudent Response," *Pakistan Observer*, n.d, accessed January 25, 2023, https://pakobserver.net/indian-journalists-military-experts-hailing-pakistan-for-prudent-response/

64 Web Desk, "Indian Journalists Criticise Missile Fiasco by Nuclear-Armed India," *News 360,* March 12, 2022, accessed January 25, 2023, https://news360.tv/en/latest/indian-journalists-criticise-missile-fiasco-by-a-nuclear-armed-india/
Anwar Iqbal Syed | Baqir Sajjad, "Pakistan Demands Joint Probe into Indian Missile Fiasco," *DAWN.COM*, March 13, 2022, accessed January 24, 2023, https://www.dawn.com/news/1679725

65 DNA, "Indian Journalists, Military Experts Hailing Pakistan for Prudent Response: BBC," *Islamabad Post*, March 13, 2022, accessed January 25, 2023, https://islamabadpost.com.pk/indian-journalists-military-experts-hailing-pakistan-for-prudent-response-bbc/

66 Naveed Siddiqui, "Navy Intercepts Indian Submarine, Foils Attempt to Enter Pakistani Waters: ISPR," *DAWN.COM*, March 3, 2022, accessed January 24, 2023, https://www.dawn.com/news/1678075

67 Editorial, "BrahMos Enquiry," *Pakistan Today,* March 13, 2022, accessed January 25, 2023, https://www.pakistantoday.com.pk/2022/03/13/brahmos-enquiry/

68 Editorial, "Accidental Missile Launch," *Pakistan Today*, March 15, 2022, accessed January January 25, 2023, https://www.pakistantoday.com.pk/2022/03/15/accidental-missile-launch/

69 Attiya Munawer, "India's Mistake Can Start a War," *Pakistan Today*, March 17, 2022, accessed January 26, 2023, https://www.pakistantoday.com.pk/2022/03/17/indias-mistake-can-start-a-war/

70 Makhdoom Din, Imran Khan, and Gen Bajwa, "Team Having Report on Indian Brahmos Missile Pakistan|," March 24, 2022, accessed January 26, 2023, https://www.youtube.com/watch?v=FjZBc_-Opx0

71 "Hayder, Haroon. Top Indian Air Force Officials Under Investigation for Firing Missile into Pakistan," *ProPakistani,* March 25, 2022, accessed January 28, 2023, Top Indian Air Force Officials Under Investigation for Firing Missile into Pakistan (propakistani.pk)

72 News Report, "Philippines Questions Indian Missile Technology," *The News*, April 7, 2022, accessed January 28 2023, https://www.thenews.com.pk/print/948086-philippines-questions-indian-missile-technology

73 Reuters, "India Sacks 3 for Accidentally Launching Missile into Pakistan in March," *DAWN.COM*, August 23, 2022, accessed January 28, 2023, https://www.dawn.com/news/1706415
Kamran Yousaf, "India Fires Three Officers over Missile Fiasco," *The Express Tribune*, August 23, 2022, https://tribune.com.pk/story/2372740/india-fires-three-officers-over-missile-fiasco

74 Web Desk, "Brahmos Missile Misfire: Pakistan Rejects Closure of Inquiry by India," *ARY NEWS,* August 24, 2022, Accessed January 28, 2023 https://arynews.tv/brahmos-missile-misfire-pakistan-rejects-closure-of-inquiry-by-india/

75 Editorial, "Indian Missile Incident," *DAWN.COM*, August 26, 2022, accessed January 28, 2023, https://www.dawn.com/news/1706864

76 Naveed Siddiqui, "Reports on BrahMos Missile Incident 'Disingenuous Attempt' to Absolve India of Its 'Irresponsible Nuclear Behaviour': FO," *DAWN.COM*, November 16, 2022, accessed January 28, 2023, https://www.dawn.com/news/1721285

77 Editorial, "India Exposed," *Pakistan Observer*, November 18, 2022, accessed January 28, 2023, https://pakobserver.net/india-exposed/

3

THE STRATEGIC COMMUNITIES WEIGH IN

The aftermath of the missile accident witnessed a major engagement by the Indian and the Pakistani strategic communities who analyzed the accident at multiple levels including the type of missile fired, the efficacy of India's command and control system, the effectiveness of Pakistan's air defense system and the implications of the accident for South Asian security in general and India-Pakistan relations in particular.

The chapter's first half begins by focusing on the Indian strategic community and the issues it discussed and debated. These issues include the rarity of the type of accident involving accidental fire of a missile by a nuclear armed state into another,[1] the timing and efficacy of the Indian state's "strategic communication"[2] in the aftermath of the accident and the questionable claim by the Indian state of the accident resulting from a technical glitch.[3] The section also looks at the Indian strategic community's focus on the launch protocol for firing of cruise missiles and the implications of its accidental firing for the credibility of India's command-and-control system,[4] the absence of "two-person launch protocol" in the BrahMos Air-Launched Cruise Missile (ALCM)[5] (although the launched missile has been claimed to be a surface launched version)[6] and whether the missile accident was due to technical or human error.[7] The section ends with a summary of the community's focus on the "sorry state" of bilateral crisis management mechanisms between India and Pakistan,[8] the prospect of growing pressure from the US non-proliferation lobby[9] and other technical and security policy issues.

The second half of the chapter looks at the issues focused on by the Pakistani strategic community such as deficiencies in Indian nuclear safety, command and control, the ostensibly tepid reaction of the international community to the accident,[10] whether the missile launch was a "technical

DOI: 10.4324/9781003543909-4

accident or deliberate design,"[11] the credibility of the BrahMos missile system,[12] the 48-hour delay in India's acknowledgement of responsibility for the accident and the non-utilisation of the "army-to-army hotline."[13] The chapter also summarizes additional important issues discussed by the strategic community such as inadequacies in Pakistan's air defense system for not being able to shoot down the errant missile,[14] poor training standards for maintenance and operating personnel handling India's missile systems,[15] the need to expand the existing pre-notification agreement for ballistic missiles to also include cruise missiles[16] and other issues.

Both the Indian and Pakistani strategic communities demonstrated significant diversity in views and this chapter undertakes an extensive review of the documentary record to capture the various positions, convergences and divergences within both strategic communities regarding the missile accident.

The Indian Strategic Community—Technical or Human Error?

As early as 12 March 2022, the strategic community in India had begun to speculate on the causes of the accident with retired Air Marshall V.K. Bhatia being quoted in *The Wire* on 12 March agreeing with the Pakistani DG-ISPR's claim that a supersonic missile had veered off from its intended path from Sirsa to its destination in the Mahajan Field Firing Range in Rajasthan. Bhatia also speculated that the abrupt change in the missile's trajectory was due to the uneven burning of its solid propellant and that India had immediately informed Pakistan via the DGMO hotline after the incident.

> In all likelihood, the IAF was conducting a routine BrahMos training exercise on March 9 by firing the cruise missile from Sirsa to the Mahajan Field Firing Range in Rajasthan some 226 km distant," military analyst Air Marshal V K "Jimmy" Bhatia (retired) told *The Wire*. But for some inexplicable reason, it went "freaky" in what was a million-in-one chance, he stated in an endeavour to try and evaluate what could possibly have gone awry with the missile. Bhatia averred that after travelling some 100km towards its objective in Rajasthan, the missile abruptly lost control, turning into a "twirler", due possibly to the uneven burning of its solid propellant. This, in turn, Bhatia conjectured had adversely impacted its on-board guidance system into which all the target coordinates had been toggled, propelling it on a totally different north-westerly direction towards Pakistan. The former fighter pilot is also of the view that soon after realising the mishap India had informed Pakistan of the mishap via the "hotline" operated by the respective Directors General of Military Operations, but this could not be independently confirmed.[17]

Brigadier Kuldip Singh (retired), writing in *The Quint* on 12 March dismissed the "outlandish" theories on social media regarding the supersonic object that violated Pakistan's airspace, and claimed that the object was indeed a BrahMos missile. Although he was critical of India's launch protocols, Singh also brought up examples of accidents worldwide in the handling of nuclear weapons to claim that human error was always a possibility when handling "strategic assets."

> From the tracking data (particularly the speed, altitude reached) released by Pakistan's DG ISPR and the pictures available, it seems fairly certain that the object was not:
>
> A ballistic missile, as Short-Range Ballistic Missiles or Tactical Ballistic Missile (range up to 300 km) typically rise to an altitude of 80–100km before re-entry;
>
> A Hypersonic Glide Vehicle, whose "booster" rocket, too, rises into near-space, with the glide vehicle descending at hypersonic speeds, ie, in excess of Mach 5
>
> India's under-development 'Hypersonic Technology Demonstrator Vehicle'/Mach-7 BrahMos-II. This is propelled to high speeds initially using a small rocket, and is then powered to the target by a supersonic-combustion-ram-jet ('scramjet') for hypersonic flight.
>
> However, it seems, particularly from pictures, that it is India's supersonic BrahMos cruise missile. Perhaps the latter was being updated with targeting data, or its systems were being tested for full-functionality, and it got fired inadvertently. The BrahMos utilises a combination of inertial navigation system (INS) and satellite-based global positioning system against land targets – and likely explains the initial track towards MFFR (Mahajan Field Firing Range, used as a waypoint) and later towards Pakistan…While this incident shows India's handling of its strategic weapons and launch protocols in poor light, the fact remains that the world over, such strategic assets are handled by humans, who are often as fallible as the next person.[18]

Bharat Karnad, a veteran Indian strategic affairs expert and Emeritus Professor of National Security Studies at the Centre for Policy Research (CPR), a New Delhi-based think-tank alleged in a 12 March post that the missile misfire indicated Pakistani intelligence infiltration of the personnel manning the Indian missile battery. He also worried that the failure of "interlocks" on the BrahMos missile would mean the need to replace such mechanisms on India's ballistic missiles. Finally, Karnad also called for a thorough probe to fix responsibility and to hand out exemplary punishment to the culprits as a deterring example.

> At a minimum, this incident suggests criminal laxity by the military personnel manning the concerned BrahMos missile battery. And at a maximum, that some of the Indian missileers were suborned by the Pakistanis.... It is in Sargodha Central's interest to claim that it saw the missile suddenly veer off mid-flight in a different direction than the one intended in order to shield the compromised/paid off Indians complicit in the act of getting a BrahMos to Pakistan-China. The first thing for the Court of Inquiry (CoI) that's been set up to do is to disregard, with extreme prejudice, the Pakistan army's account of it....Still, it is something of a revelation that the series of electronic interlocks built into the triggering system of the missile can be bypassed....Would it be very wrong, in the event, to assume that the strategic nuclear warheaded Agni missiles are under a similar locking system with the final authority to fire being easily side-stepped by the unit in the field?...Re-engineering the interlocks system in the missile triggering mechanism, one in which the option of manual over-ride at the local level is denied without an absolute final authorization, is an urgent necessity....The air force officers and other ranks up and down the line and party in any way to this mishap will have to be held responsible. Those directly implicated will need to be drummed out of service.[19]

In an interview with *Rediff.com* on 14 March, Karnad cast doubt on the credibility of the Defence Ministry's explanation that a "technical malfunction" had resulted in the accidental launch. He contended that the accidentally launched BrahMos cruise missile was nuclear-capable with interchangeable conventional and nuclear warheads. There was an elaborate sequence of steps leading up to a final authorization before its launch.

Interviewer—*The defence ministry seems to have landed with egg on its face.*

Karnad—*A whole barnyard full of eggs, in fact. This is quite shocking and simply cannot be credibly explained away by referring to a "technical glitch". The triggering mechanism is a hardy piece of work including a firing sequence and a final authorisation. How this process was obviated is a mystery.*

Interviewer—In your view, could this have been a BrahMos cruise missile possessing nuclear capability?

Karnad—*The BrahMos missile has interchangeable warheads and can carry both conventional and nuclear weapons. But most forward-deployed Indian cruise missiles are conventionally armed.*[20]

Military Hotline Not Used, Missile Not Shot Down

Ambassador Rakesh Sood, former Special Envoy of the Prime Minister for Disarmament and Non-proliferation Issues[21] and a Distinguished Fellow at the New Delhi-based Observer Research Foundation (ORF), analyzed that India did not use the military hotline to inform Pakistan about the accidental launch.

> The Director Generals of Military Operations' hotline is intended to provide a secure and fast channel of communication to deal with any incident that has a crisis potential and thereby avert any undue fallout," said Ambassador Rakesh Sood, an expert on nuclear issues. "Clearly, in this case, it doesn't appear to have been used. Consequently, we are witnessing the unseemly exchanges that have taken place in recent days," said Ambassador Sood.[22]

In a *Times of India* podcast on 14 March, editor and podcaster Meenal Baghel conversed at length with her two guests regarding the events surrounding the accident and its implications for regional security. Manoj Joshi, a distinguished fellow at the New Delhi-based Observer Research Foundation (ORF) and a member of the Indian government's Task Force on National Security[23] opined that the missile "definitely looks like a BrahMos." He added that he was unaware as to whether BrahMos missiles were tested at the Mahajan Field Firing Range or at sea in Gopalpur as in the past. He expressed surprise that the missile was tested over land.

Joshi explained that missiles are produced in batches and one of them is then tested and that "some component has obviously gone haywire" which has led to the missile accident. Taking note of the Pakistani claim that the missile was detected at 40,000 feet, Joshi explained that the BrahMos was not usually supposed to fly at that height as it was originally designed as an anti-ship missile which did not require flying at such heights. He speculated that this high altitude flying by the missile too, could have been part of the malfunction. Explaining that the BrahMos is widely distributed among Indian armed forces as it is capable of penetrating enemy defenses because of its speed and as a shock weapon, Joshi reckoned that the confusion within the army bureaucracy regarding the position and final location of the missile could have resulted in the Indian silence for the first 2 days of the crisis. Joshi also elucidated that the Court of Inquiry (CoI) process would involve a senior officer being assigned to investigate and the officers commanding the missile unit would be asked to depose. The CoI would also look at the forensic evidence collected. The other missiles from that particular batch would be tested at sea to check for any technical defect. In light of Pakistan's repeated demands for a joint probe, Joshi doubted whether Pakistan could play a contributing role in any joint investigation unless it was keen on returning parts of the crashed missile which can be used in the Indian investigation. He concluded by averring that India and Pakistan have for too long postponed a serious discussion of nuclear weapons and the BrahMos misfire could be a wakeup call for this larger discussion.

Sushant Singh, Senior Fellow at the New Delhi-based Centre for Policy Research, pointed out (based on the Defence Ministry's statement) that the missile seemed to not have been launched accidentally during a test but during an inspection. He questioned whether Pakistan needs to discuss a revision of its Standard Operating Procedures (SOPs) for allowing the missile to land, especially given that the Pakistani Air Defense System had no way of knowing whether the incoming missile was armed or unarmed and if the warhead on an armed missile was conventional or nuclear. He seemed puzzled by the Pakistani inaction given the oft-repeated Pakistani fear of Indian missile systems falling into the hands of religious extremists. Singh also speculated that Pakistan perhaps might have intercepted the missile had it flown towards a nuclear facility or military base or populated area. Since the BrahMos was not a ballistic missile, India was not bound to pre-notify Pakistan about the same under the 2005 Agreement on Pre-Notification of Flight Testing of Ballistic Missiles.

Singh also warned about the compressed response time of 6 minutes that India and Pakistan had to respond to an accidental launch due to geographic proximity as compared to 30 minutes for the US and Soviet Union and how this increased the chances of escalation due to a hasty and disproportionate response. He analyzed that the statements coming out of Pakistan especially from National Security Adviser Moeed Yousuf were Pakistan's attempts to re-hyphenate itself with India, a hyphenation that India had escaped from the period of the Indo-US nuclear deal.

The motive for such re-hyphenation was for international non-proliferation pressure to be brought on India. Singh also expressed puzzlement as to why India did not use the DGMO hotline to inform Pakistan after the accident and called for a transparent investigation to affix responsibility thereby reclaiming India's global credibility as a responsible state, an image dented by the missile accident. This was important as India's responsible state image had earlier helped it avail exemptions such as the Indo-US nuclear deal. He also called on the Indian government to review its SOPs and communication systems and to ascertain whether the personnel being chosen to command missile units were undergoing the necessary political, ideological and intelligence vetting.[24] Both Joshi and Singh concurred that the unfilled post of the Chief of Defense Staff in the aftermath of General Bipin Rawat's death in a helicopter crash had implications for the crisis management and communication after the missile accident. While the occupant of this post would not have been able to prevent the missile accident, they could have better managed the channels of communication with the political establishment to better mitigate its consequences.[25]

Tara Kartha, Distinguished Fellow at the Institute of Peace and Conflict Studies, New Delhi with 17 years of experience at India's National Security

Secretariat and special charge of the National Security Advisory Board, framed the missile accident as an "opportunity" for India and Pakistan to dialogue in order to ensure "stability in conflict."[26] She also pointed out that the accidentally launched missile had nonetheless managed to penetrate Pakistani air defenses.

> India's Draft Nuclear Doctrine follows a policy of "minimum and credible deterrence", which follows an equally unambiguous "no first-use" policy (NFU). There is no question of India firing any kind of missile at Pakistan, especially in a situation where there are no high tensions prevailing at all. Pakistan's unstated doctrine is based on first use, and includes tactical nuclear weapons, seen as dangerously destabilising given unknown command and control over them. But in this case, Pakistan's armed forces displayed a remarkable degree of maturity in not jumping to the worst conclusions....At a second level is the simple fact that a BrahMos cruise missile got in through Pakistani air defences. That's the bottom line.[27]

Missile as Modified ISRO Rocket, Launch a Deliberate Act, Praise for Pakistani Restraint

Writing in *Daily News and Analysis* (DNA), an online news outlet, Indian navy veteran Manan Bhatt contextualized the missile entering Pakistani airspace as the latest iteration of a pattern of "touch and go" Combat Air Patrols (CAP) flown by the Indian Air Force to "spook off" Pakistan's air defenses and to test its reaction times. Manan also made some highly specific claims regarding the errant missile's nuclear warhead not being activated at the time of launch, its impact in Mian Channu being part of a deliberate targeting strategy and the commander of the Pakistani air defense network almost being terminated for not shooting down the Indian missile.

> Unconfirmed reports originating from India suggest that the "missile" was actually a modified version of the ISRO (GSLV MKIII – MI) rocket which had been mistakenly shipped to Sirsa...Unofficial reports are also doing the rounds that though the "flying object" did have a tactical nuclear warhead (less than 1 KT) mated to it, the activation codes could not be located during maintenance period, nor was the 'self-destruct code' available at hand as the same had been kept under lock and key for safe keeping....Initial reports suggest about a major shake-up in PAK's air defence establishment, citing incompetency on their part. A Qatar-based international news channel reported about sacking of Pakistan's top Air Defence Commander after the incident. In addition, the Vice Chief of the Pakistan Air Force and one more Air Marshal have been asked to put in their papers for voluntary retirement....Pakistan has commenced a major

> audit and RF frequency change of all passive and active radar emitters as well as ECM, ECCM and allied EW measures on emergent basis.... Pakistan's defence establishment is averse to using missile defences, unless the target is GHQ. Can cash-strapped Pakistan even afford to test fire the LY-80 air defence missiles costing a whopping 6.27 million apiece?
>
> Pakistan's Chinese-made LY-80/HQ-16 air defense – Surface to Air Missiles batteries and IBIS-150 air defence radars didn't track or intercept India's Balakot Air Strike on February 26, 2019 and the incoming misfired missile on March 09, 2022...Pakistan claimed there was no warhead in the missile. But visuals shared by ISPR and open source videos of the crash site reveal the extant of the damage which shattered the entire structure, collapsing the walls. A warhead bursts over pressure can only cause such widespread damage.[28]

In a 14 March telecast of the show "No Spin" presided over by anchor Nidhi Razdhan of the left-leaning NDTV, Lieutenant General (retired) Satish Dua, former Chief of the Integrated Defence Staff, expressed surprise at the incident, even admitting that he initially thought the misfire to be a hoax. He averred that the CoI's findings would help ascertain the reasons for the accident. Dua expressed relief that Pakistan did not retaliate as such retaliation could have led to rapid escalation. Commenting on the firing sequence of the missile, Dua explained that there are multiple mechanical and software-driven electronic locks which could be bypassed due to a technical malfunction or due to human error with both being serious.

He also speculated that either Pakistan was not able to detect the incoming missile or its air defense systems were not on full alert due to recent stability in the India-Pakistan relationship which reduced the probability of an airspace violation. Dua also compared the missile accident to the actions of Soviet Lietenant Colonel Stanislav Petrov in disregarding the report of early-warning systems of an incoming US nuclear missile attack in 1983 as a false alarm and his subsequent disregarding of Soviet military protocol which called for a retaliation. Dua was also categorical that there was no requirement for or possibility of a joint probe as the accident was a military matter and that its findings would not be made public. He however admitted that the accident had exposed various inadequacies in both India and Pakistan. Dua concluded that there was an urgent need for more official lines of communication between India and Pakistan with military to military communication being especially important.

Tara Kartha, the other guest from the Indian strategic community on the show praised the Pakistani DG-ISPR's statement on the missile accident as considerate and mature, especially given that the missile crossed three Pakistani districts. She contrasted its restrained response to that of the Pakistani Foreign Ministry and NSA which she characterized as being more aggressive. She also questioned as to why Pakistan did not shoot at and shoot down the missile if it did track it since Pakistan has a comprehensive, layered, integrated defense network which includes the Chinese-supplied HQ-9, claiming to intercept a missile at 100 km in one shot. Contending that CoI's (Court of Inquiry) worldwide were not open to public scrutiny, Ms Kartha expressed satisfaction over the actions of both countries in the days following the accident and concluded by calling for cruise missiles to also be brought under the ballistic missile notification agreement (signed in 2005).[29]

BrahMos as Air Launched Cruise Missile (ALCM), Two-Man versus One-Man Protocol

Retired colonel and strategic affairs columnist, Ajai Shukla, claimed in a 15 March piece in *The Business Standard* that the misfired BrahMos cruise missile was an Air Launched Cruise Missile (ALCM) and could not have been mistaken by Pakistani air defense for a ballistic missile. Shukla also averred that the failure of protocol in the case of misfired BrahMos missile could not be equated to a failure of India's ballistic missile launch protocols as the former had a one-man protocol as opposed to a more secure two-man protocol for the latter.

> It is a tenet of India's nuclear launch posture that strategic weapons – commonly referred to as "nukes", or nuclear-tipped missiles – are invariably delivered via ballistic missiles, never via cruise missiles. So there was little prospect of a cruise missile, fired from Sirsa or elsewhere, spooking Pakistan into setting off a nuclear exchange. Since the cruise missile had been launched from near Sirsa, it would have almost certainly been identified by Pakistan's air defence network as a BrahMos air launched cruise missile (ALCM). That is because Sirsa is home to a major air force base, but not to any army BrahMos units, which are restricted to the strike corps. Thirdly, the violation of Indian launch protocols is not a matter of grave concern in an accidental BrahMos ALCM launch. Unlike in the launch of a strategic, nuclear-tipped ballistic missile, where a two-person launch protocol is mandatory, this is not so in the case of a BrahMos ALCM launched from a Sukhoi-30MKI fighter. In the latter case, once armament is loaded and activated, there is no requirement for a two-person protocol to protect its launch. Since there is always the possibility of one pilot being rendered casualty and, therefore, unable to play his role in triggering a launch sequence, a single-person firing protocol is regarded as sufficient for an ALCM.[30]

Dr Rajeshwari Rajagopalan, Director of the Centre for Security, Strategy and Technology at the Observer Research Foundation and former Assistant

Director of the National Security Council Secretariat (2003-2007),[31] opined in a 15 March interview with veteran journalist Karan Thapar of *The Wire* that the missile accident was a serious issue and requires an investigation. She also averred that it was not clear whether Pakistan actually detected the missile and that it was being assumed that Pakistan did not react even after it detected the missile. She also pointed out that the Indian claim, being made through anonymous sources in certain media outlets that Pakistan was immediately informed after the accident did not specify the exact time frame in which such a warning was given. On being pressed by Thapar as to whether there was any basis to completely disprove the Pakistani claim that India did not use the DGMO hotline, Pillai admitted that her understanding of an early Indian communication to Pakistan in the immediate aftermath of the missile accident was based only on media sources.

She was clear that, whether the accident occurred during routine maintenance or whether it occurred during an inspection at an IAF base, the accident raised serious questions about India's protocols and practices. More specifically, questions would also arise about whether there was a lack of adequate protocols which caused the accident or whether there was a lack of compliance of existing protocols with the latter being a bigger problem. Rajagopalan also analyzed that long-range missiles such as the BrahMos would definitely have a self-destruct mechanism and pointed to a lack of information in the public as to why it was not activated.

She hypothesized that perhaps the on-ground personnel commanding the missile unit were waiting for orders from higher authorities to activate the mechanism or maybe they activated it and it failed. She also worried about additional possibilities whereby the personnel commanding the missile unit were not aware of how to activate the mechanism or by the time they realized that the missile had strayed from its intended path, it was too late to activate the mechanism. She informed that many missiles have a feature wherein the self-destruct mechanism cannot be activated once the missile is out of range. She then speculated that the focus of the CoI would be on the missile's firing and the reasons for the same and that it was not necessary or prudent for the government to release information about the missile's flight path as this would benefit Pakistan.

She explained how official confirmation of the missile's flight path would provide Pakistan with the info regarding the gaps in its air defense system and would also provide Pakistan with vital information of the usual flight paths taken by Indian missiles thereby allowing it to strengthen its air defenses along those flight paths. She also expressed puzzlement over reports in the media quoting unnamed government sources that the missile followed a pre-determined flight path but did not hit the pre-determined target due to certain unstated factors. She contended that while the aforementioned unstated factors needed to be known, there was also the possibility that the missile neither had a pre-determined flight path nor a target but landed where it did because it ran out of fuel. An additional possibility was also that there was indeed a pre-determined flight path and a target for the missile from a certain origin point (in South India) and as the missile was fired from a different origin point, the target and flight path did not apply. The aforementioned possibility was taken by her as a sign of incompetence and poor training at institutions which could be redressed through regular simulation exercises and contingency planning for similar situations, focusing in particular on the chain-of-command's crisis communication efficacy in an accident's immediate aftermath.

She also advised that the Indian government not react defensively but approach the issue from a point of strength with a frank acknowledgement of responsibility, a thorough CoI completed in a reasonable timeframe and a clear articulation of the steps that would be taken to prevent a recurrence. She suggested that the Indian government brief its allies and partners in private on the details of the accident, the actions taken and the protocols and practices that have been revised. While cautioning that India's response to the accident would affect its image as a responsible state, she also comforted that the responsible state image was based on a broader set of considerations such as India's No-First-Use policy and refusal to proliferate technology to other countries.[32]

Risk to South Asian Stability—Armed Forces Believe Human Error Caused Accident

Dr Happymon Jacob, Associate Professor of Diplomacy and Disarmament at the School of International Studies, Jawaharlal Nehru University (JNU), highlighted the poorly developed bilateral crisis management mechanisms between India and Pakistan and the risk this poses to South Asian stability.

> The accidental firing of an Indian missile into Pakistan on March 9 calls for serious introspection by the two nuclear-armed adversaries about the perils of living under the shadow of nuclear weapons. The unfortunate incident also casts a shadow on the standards of the storage, maintenance, the handling and even the engineering of high-technology weapon systems in India. But, more pertinently, the incident highlights the sorry state of bilateral mechanisms for crisis management between the two nuclear adversaries where there is a missile flight time of barely a few minutes. The Pakistani response to the accidental firing of the missile was a balanced one especially when handled by the Pakistan Army on March 11. While New Delhi maintained a silence over the issue until it was brought up on March 11, the Indian response was also far from denial.[33]

Manvendra Singh, Congress party leader and Editor-in-Chief of *Defence & Security Alert*, criticized the IAF for its laxity and lamented that it had eroded India's image of being a responsible power in a 16 March article for *The Print*. Singh also highlighted the sentiment within the armed forces which was dismissive of the technical malfunction explanation initially advanced by the Defence Ministry.

> After causing international embarrassment of the highest order, the government and its "sources" have changed their position on the launch of the missile from the plausible to the fantastic. None of the positions taken does anything to redeem the damage done to India's current image of being a technologically irresponsible military power....Military memes, circulated within the armed forces community, mocked the "technical malfunction led to the accidental firing" claim with a brutality that is unrepeatable here. For the armed forces know, above all, the gravity of any accidental firing. Even the inadvertent firing of a small arm is treated with the seriousness that it deserves, for the smallest bullet can take a life, accidentally. So within the military community, the BrahMos incident is being regarded as grave error. It is high time the Air Force takes the matter to its logical conclusion. As it is certainly not the first such in the recent past....Such lapses when overlooked come back to haunt the Service, and now the Air Force has a lot to answer who and why one of their own pressed the launch button of supersonic cruise missile.[34]

In her 16 March show on *Mojo Story*, a Delhi-based news video publisher, veteran journalist Burkha Dutt, discussed the accident and its implications with Lieutenant General and former Chief of the Army, Raj Kadyan, and India's former high commissioner to Pakistan, TCA Raghavan. Kadyan agreed with Dutt's observation that both India and Pakistan were able to deal with the crisis in a mature and responsible manner. He also cautioned that speculation would be rife regarding the causes of the accident until the CoI's report was expeditiously made public. He analyzed that whether the COI discovers a technical malfunction or a human error as the cause of the accident, India will not come out looking as a responsible state capable of dealing flawlessly with its sensitive strategic assets with the missile gaffe having negative implications for India's image. General Kadyan also questioned why the missile wasn't shot down by Pakistani air defense despite being in its airspace for three-and-a-half minutes. He also pointed to speculation that Pakistan had sacked its general responsible for air defense which if true suggested that it wanted to intercept the missile but couldn't.

High Commissioner Raghavan also agreed with General Kadyan that the accident was handled well by both countries, especially Pakistan which restricted its response to strong statements. He also situated the missile accident as one among a whole range of issues in the India-Pakistan relationship with escalatory potential. Commenting on the *Bloomberg* report on Pakistan planning and then aborting a retaliatory strike, Raghavan averred that reports of such a speculative nature were bound to be published in the aftermath of the crisis/accident. He analyzed that the militaries on both sides had a good understanding of each other's capabilities which led to the restrained reaction and called for a broader political understanding and long-term engagement to rebuild trust in the relationship.[35]

Error in Firing Sequence of Missile—Major Violation of Standard Operating Procedures

D. Raghunandan, Director of the Centre for Technology and Development of the Delhi Science Forum raised several questions regarding the firing sequence leading up to the missile accident in an article for *Newsclick* on 17th March. He also criticized the government for its tendency to shield the armed forces from accountability for their professional performance.

> The first big question obviously is, how did the missile get "accidentally fired" during maintenance? The Statement by the Defence Minister in Parliament stated that the Standard Operating Procedures (SOPs) of missile maintenance would be reviewed, indicating that the SOPs were not observed or supervised correctly, and therefore perhaps require modification....First, it should be noted that the warhead was clearly de-mated from the missile, which is good if long-standing safety practice involving nuclear weapons, and reduces chances of impulsive or sudden launches, even though it is not known whether this particular BrahMos missile was nuclear-tipped or not. Secondly, however, why was the firing mechanism kept on during maintenance? Usually, the firing and the direction-setting mechanisms are fully encrypted and require proper coded de-activation and re-activation. Some newspapers have reported that the codes were not available at the spot. While this could not be independently verified, this aspect of the SOPs is clearly worth looking at carefully, given also the erratic trajectory of the missile once let loose. It is also not known whether this missile, or batch of missiles, had built-in self-destruct mechanisms or not, and, if they did, why they could not be activated.....For us here in India, it is important that the government conducts a clean, impartial and brutally honest inquiry, and not to hide findings behind the convenient cloak of national security, ultra-nationalism, or arguments that the armed forces are beyond scrutiny or criticism.[36]

Particularly important were the views expressed by former Navy Chief Admiral Arun Prakash in a 17 March interview in *The Telegraph*. Prakash worried that both suspected causes (human error versus technical error) carried serious implications for the command-and-control system and missile design processes, respectively. As a former Navy insider and member of its top-brass, Prakash seemed to put less trust in the one-man versus two-man activation protocol that retired Colonel Ajai Shukla had earlier explained to assure that it would be lot more difficult to accidentally fire a ballistic missile as compared to a cruise missile such as the BrahMos.

Interviewer: How can a missile be accidentally fired?

Admiral Prakash: It sounds very strange. Missiles, even most basic weapons, have at least two-three safety systems built in them. This (missile) is a strategic weapon system. So I doubt it's possible to fire it by accident. Obviously it's a human error or a technological issue. It is even more disturbing if there was some problem in the design. If it's not a design issue then it's a human error, which means procedures are not well laid down....There is an international rule that a missile should never be launched accidentally but it should always be launched when duly authorised. There is no question that one can fire a strategic weapon by accident or mistake....This is the basis of all command-and-control structures worldwide. So that rule has obviously been violated because it has gone off without due authorisation, mistake or accident—whatever you call it. That's not a good sign for a command-and-control system... Finally, this was a cruise missile, it could have happened with the ballistic missile also and in that case the consequence could have been far more serious.[37]

Former Air Vice-Marshall Kapil Kak also gave his perspective as a former insider on the steps that lead to the firing of a missile in the same aforementioned 17 March interview. Kak also characterized the accidental launch as a major violation of launch-related standard operating procedures (SOPs).

Interviewer: What are the steps up to the firing of a missile?

Air Vice-Marshal Kak: A missile has a launch platform. The missile should be mounted on the launch platform and all the electrical circuitry has to be energised. All the mechanical locks need to be unlocked and ensure that all the propellants are there, as without these the missile cannot take off. Electronic circuitry takes place and once you press the button the missile goes off.

Interviewer: Is it possible for a missile to be fired accidentally.

Air Vice-Marshal Kak: This is certainly in the realm of possibility. If there is a slight human error in disregarding certain launch protocols, in disregarding certain electronic and mechanical locks, it can happen. Accidental firing can take place, it's possible.

Interviewer: What does this say of the command-and-control system?

Air Vice-Marshal Kak: I feel that it ("accidental" missile firing) should not have happened as it speaks a lot about the command-and-control system. This is a serious breach of launch discipline and protocol.[38]

Without pulling his punches, Kak also criticized the Indian government's delayed response and terse communication as the result of centralized political control

Interviewer: How do you see the government's response?

Air Vice-Marshal Kak: The Indian government's communication strategy on this has been below par. It is because of political control as it is overly centralised. The accident happened around 7 pm on March 9 and I think we realised soon enough that we had goofed up our missile. I think we would have realised within five minutes, the Indian Air Force chief would have got the information that a missile had been fired by mistake. In the next five minutes, the Prime Minister should have authorised issuance of a statement internationally that a missile was fired and we are examining the situation and we will ensure it does not happen in future....In a conflict situation it could prove critical. The other side could feel that a nuclear missile has been launched.[39]

An article in *The Print* on 18th March reproduced the Asian News International (ANI) report of a Pakistani missile launched from Pakistan's testing range in Sindh and crashing shortly after in Thana Bula Khan in Sindh. The article relayed speculation by Pakistani media that the failed test was Pakistan's reaction to India's accidental BrahMos launch.

> Islamabad [Pakistan], March 18 (ANI): The residents of Jamshoro in Pakistan's Sindh province noticed an unidentified flying object falling from the sky with a tail of smoke on Thursday. The object clearly resembled a rocket or a missile. According to information gathered from social media, the object was a missile, fired by Pakistan from its test range at Sindh. The test was scheduled at 11 am which was postponed by an hour owing to a fault in the TEL (Transporter Erector Launcher). It was finally conducted at noon. However, seconds after the launch the missile was seen descending from the path, clearly falling short of its desired target and crashed near Thana Bula Khan in Sindh. Although few Pakistan news channels covered the incident, the authorities in the country have remained silent. According to social media accounts in Pakistan, the local administration, however, refuted any such claim saying that it was a regular Mortar Tracer Round that was fired from the nearby range.

> It is unlikely, however, that a mortar with a maximum range of 5 km will have a tracer projectile rise that high....Pakistani news agency the *Conflict News Pakistan* reported that Pakistan might have test-fired a missile in response to a previous missile mistakenly fired from India..... In the NOTAM, the details of the testing schedules are specified in which the dates 17th and 18th March are mentioned.[40]

Three Possible Reasons for Missile Accident, Uncertainty over Indian Use of Hotline

Retired Air Marshal Anil Chopra, head of the Centre for Air Power Studies (CAPS), a New Delhi-based think-tank, analyzed the possible causes for the missile misfire in a 19 March interview with journalist Krishn Kaushik of the *Indian Express*. Chopra speculated on three main causal factors: 1) incorrect target coordinates, 2) cyber jamming of the missile in mid-flight, adversely affecting the coordinates and 3) corruption of target coordinates fed into the missile.

> Retired Air Marshal Anil Chopra, who heads the Centre for Air Power Studies (CAPS) think tank in New Delhi, said there are very few reasons for a missile to change its direction thus....For a cruise missile, "you have to give target coordinates" when fired from the ground, and "after that she is on her own", he said. There are also some missiles for which the coordinates can be updated in flight. "First thing could be that the coordinates are not correct. But in this particular case, the missile has gone in a particular direction, and then turned. If it was wrong coordinates, it should have gone straight there. Because...normally it will only turn in the last stages....Another possibility "is if somebody was to jam the missile while in flight, by some cyber means—I am just conjecturing. Then the whole coordinates get affected, changing the direction." He noted that normally, "a missile like that has a destruct fuse, which means from the ground you should be able to destroy it in flight if you think it has gone haywire. We do not know: was it loss of contact altogether, that you could not activate self-destroy?"...What can also cause a malfunction is "if the target data that has been fed into the missile gets corrupted, then takes a different direction all together." Had the missile crashed, "then we know some controls had failed...She had flown straight, then turned, then flown straight...Considering that, if she has not done any fancy manoeuvres, has sometime during flight the destination got corrupted?" He noted that cyber intervention is in neither country's interest. "Maybe it's a one-time failure inside, of one digit, in the coordinates."[41]

Dr. Rajeswari Pillai Rajagopalan, Director of the Centre for Security, Strategy and Technology (CSST) at the New Delhi-based Observer

Research Foundation and Assistant Director at the National Security Secretariat (2003–2007),[42] highlighted the key uncertainties that persisted around the missile accident and its serious ramifications for regional security. She worried in particular about the conflicting claims by Pakistan and India (through anonymous accounts appearing in the Indian media) about whether India had used the military hotline to inform Pakistan about the accidental launch.

> It is unclear from the different Indian statements, and even from the unofficial source-based reporting in Indian media, whether the missile launch took place because of technical problems, some kind of human error, or some procedural lapse. Each has a different implication. The most concerning would be the question of human error or procedural lapse. The Indian military services over the last several years have witnessed many high profile tragedies and mishaps, including a helicopter accident that led to the death of India's first Chief of Defense Staff, General Bipin Rawat, his wife, and several other officers in 2021....Equally unclear but with more dangerous implications is the question of precisely when India informed Pakistan about the accident....The hotline issue is particularly salient as Pakistan's National Security Advisor Moeed Yusuf has claimed that India did not use the hotline. India, however, states that it informed Pakistan about the errant missile well before Pakistan's public statement on March 10. It is, of course, possible that India informed Pakistan without using the hotline. The issue is important because it potentially affects how the two countries will behave and how much trust they will have regarding each other in any future crisis.[43]

Manish Tewari, Congress party leader and former Minister of Information and Broadcasting, writing in *The Asian Age* on 26 March called for an institutionalized dialogue between India and Pakistan to put in place mechanisms to prevent misunderstanding and the resulting conflict escalation. Tewari's article lumped the risk of escalation due to the BrahMos (cruise missile) misfire with that of escalation due to an accidental ballistic missile release.

> Had the Pakistanis reacted to the missile launch with a retaliatory strike of their own it would have been game and match over in three to seven minutes—that is the missile flying time between India and Pakistan. Both countries could have been nuclear toast. Given the various ominous implications of this missile launch whether due to a technical malfunction or during routine maintenance and inspection it presents a very strong case for an institutionalised dialogue between India-Pakistan on nuclear issues and a regular exchange of strategic force commanders.[44]

Keeping Conventional and Nuclear Weapons Systems Separate to Avoid Entanglement

In a 31 March online panel discussion titled "Skirting the Nuclear Threshold: Conflict De-escalation and Strategic Stability in South Asia," organized by the New Delhi-based Council for Strategic and Defence Research, Happymon Jacob, an Associate Professor of Disarmament Studies at the School of International Studies, Jawaharlal Nehru University, discussed the implications of the missile accident with two Pakistani and two Indian strategic experts.

Jacob began by noting that there exists a "cold peace" between India and Pakistan, since the ceasefire agreement of 2021. Noting that accidents involving India and Pakistan had happened in the past with the potential for similar accidents to occur in the future, Jacob commended Pakistan's ISPR in particular for handling the missile misfire in a mature and restrained manner.

Jacob also praised both sides more broadly for their mature management of the incident which prevented escalation and held it out as a precedent to be followed in case of future accidents. He averred that India should have activated the DGMO hotline to inform the Pakistanis and worried that the accident had fed into Pakistani suspicions that India was preparing for some kind of counterforce strike. He also analyzed that India's No First Use (NFU) policy seems to have reassured the international community that the missile launch was an accident and wondered as to whether the Indian NFU had sufficiently reassured Pakistan about India's lack of deliberate intentions in case of the missile accident.

Jacob also brought up the lack of a structured process of interaction with regularized exchanges between India and Pakistan, pointing to the absence of High Commissioners in both countries and absence of expert-level discussions on nuclear confidence building measures for the past seven to eight years. He warned that piecemeal confidence building measures were inadequate. Pointing out that the Indian practice of keeping conventional and nuclear systems separate had been a good one as it provided Pakistan better situational awareness during incidents such as the missile misfire, Jacob also spoke approvingly of the de-alerted and de-mated posture of Indian and Pakistani nuclear weapons. Jacob ended with an enjoinment for more regular consultation between the two countries to clarify nuclear doctrines, respective redlines, force postures and emerging threats to stability in the cyber domain.

Dr. Manpreet Sethi, Distinguished Fellow at the Centre for Air Power Studies (CAPS), New Delhi and a Senior Research Adviser at the Asia Pacific Leadership Network (APLN), termed the accident highly regrettable and noted that accidents had occurred involving the US and Russia during the

Cold War both due to human error and technical malfunction. She also noted the potential for learning lessons from accidents such as the missile misfire for the countries involved. Noting India's announcement of a high-level Court of Inquiry, Sethi analyzed that a tightening of technical, operational and organizational Standard Operating Procedures (SOPs) could be expected in the aftermath of the incident. She concurred with Jacob regarding the wisdom in India's conscious strategy of maintaining a distinction between nuclear and conventional systems (unlike China and Pakistan) thereby avoiding nuclear entanglement with dual use systems which constituted a risk reduction measure of its own kind.

Sethi appreciated the Pakistani establishment's mature response to the missile intrusion, especially the restraint shown by the ISPR in its press briefing that focused on releasing the information gathered into the public domain and did not encourage speculation. The incident's occurrence during a period of peace between the two countries also resulted in Pakistan having little reason to suspect that there was a mala fide intention on India's part.

The Indian strategic community's war games have often concluded that neither India nor Pakistan is really trigger happy and the real-time restraint shown by both countries in the aftermath of the missile incident attests to this. Sethi also pointed out that despite South Asia being referred repeatedly in Western strategic discourse as a nuclear flashpoint, there was greater restraint exercised by India and Pakistan than what is currently being displayed in Ukraine. Commenting on the angst in Pakistan that India was not sufficiently chastised over the accident by the international community, Sethi contended that Pakistan's culpability in terrorism and nuclear proliferation had lowered its international credibility. India's responsible behavior in the past had led the international community to view the missile misfire as an accident and to repose confidence in the Indian assurance that SOPs would be revised to prevent a recurrence.

Sethi also informed that the proposal for cruise missiles being brought under the 2005 ballistic missile testing agreement's pre-notification requirement has been discussed at the track two level for several years but it was the 'cold peace' between the two countries that prevented them from reaching an understanding. She identified the aforementioned proposal as a low hanging fruit given that cruise missiles were only conventionally armed in case of India while they were nuclear capable in case of Pakistan.

Sethi also called for a better utilization of military hotlines. Sethi drew attention to the uniqueness of the 2007 agreement which required India and Pakistan to notify each other in case of a nuclear accident but noted that the implementation of the agreement through relevant SOPs and points of contact still needed to be worked out. She worried that the current state of minimal to no engagement at the bilateral level was not helpful as it could lead to one-sided perceptions of the adversary, reading much more into its actions and resulting in reactions that could further widen differences.

Lt Gen D.S. Hooda, former commander of the Indian Army's Northern Command and a senior fellow at the Delhi Policy Group[45] reckoned that the Indian government had taken serious cognizance of the missile accident and that a thorough inquiry would be carried out as it concerns the safety and security of India's high-value weapons systems. He also commended both countries for showing a great deal of maturity in the way they dealt with the accident and its aftermath. Hooda brought up three major issues of importance stemming from the accident. First, there was a lack of communication between both sides despite existing hotlines between the DGMOs and Foreign Secretaries. The fact that none of these hotlines were activated by either country is really worrisome and there was a need to ascertain why this happened.

The second major issue of concern articulated by Hooda was that, the whole issue of strategic and nuclear stability cannot only be looked at only when a crisis occurs. Even if CBMs cannot be implemented due to absence of an enabling political climate, accidents can happen at any time resulting in risks to regional stability. In particular, Hooda reminded his fellow panelists of the *Bloomberg* article (covered in Chapter 2) which quoted a Pakistani official informing that Pakistan had initially considered retaliation but decided to not go through with it. The third issue of concern which Hooda flagged was the difference in the nuclear doctrines of India and Pakistan. He explained that India's NFU had at its core a perception of nuclear weapons only to be deployed to fulfill its threat of massive retaliation to deter the use of nuclear weapons on its territory. He contrasted the Indian view with that of Pakistan which viewed nuclear weapons (including tactical nuclear weapons) as a means of deterring conventional war, as part of its full-spectrum response and as part of its warfighting strategy.

Hooda characterized the Pakistani position on nuclear weapons as inherently more destabilizing. An important point made by Hooda was to contend that it would be wrong to equate command and control glitches in the BrahMos missile with that of nuclear systems, as the latter was the exclusive responsibility of the Strategic Forces Command. He stressed that India had deliberately kept its conventional systems separate from its nuclear systems with different command and controls for each.

Hooda also averred that Pakistan's demand for a joint probe was not possible practically as India would not allow a hostile country to look at its technical issues, SOPs and protocols. He also expressed doubts over whether the probe's results would be shared. Hooda concluded by emphasizing that both sides further clarify their nuclear doctrines, find ways to insulate political ties from nuclear CBMs, and agree on mechanisms to prevent cyberattacks on each other's nuclear installations and command and control facilities. He identified the 1999 Memorandum of Understanding (MOU) as a good base document to build on as the MOU was also mentioned in the preambles of the 2005 and 2007 agreements respectively.[46]

India's Post-accident Messaging, Failure of SOPs and "Second-Rung" Missile Operators

Ruhee Neog, Director of New Delhi-based Institute of Peace and Conflict Studies (IPCS) and the Managing the Atom & International Security Research Fellow at the Belfer Center, Harvard University, opined in a 29 April article that India's post-accident messaging had failed the "three tests of communication: language, timeliness and narrative control."

> the language India employed in the first statement (11 March) lends credence to the perception of complacency. Some of this language is a consequence of post-colonial South Asia's preference for the agentless passive voice. Consider the sentence "It is learnt that the missile landed in an area of Pakistan." This statement has no agent, obfuscating who "learnt", and suggesting that the missile acted by itself, i.e. malfunctioned. This phrasing might have been clever, had it not come off as a dereliction of responsibility and an arrogant trivialization. That the official notification was issued two days later compounded this perception. Timing is everything in strategic communications. So are skilled drafters....By not reacting swiftly, India conceded first mover advantage to Pakistan. A useful analogue is how the media interacts with government decision-making during crises. In the absence of clear, unified messaging, the press effectively rushes in to fill the news vacuum. Pakistan, like India, has a vested interest in shaping the political narrative in its favour. It took an opportunity where it saw one.[47]

Given the sensitivity and official secrecy around the accidental launch, there was a brief interlude in the debate until 22 August when the government's punitive actions based on the conclusions of the CoI became clear. Writing in *The Quint* on 25 August, K.P. Sanjeev Kumar, a former Navy experimental test pilot, reckoned that based on the mechanical and electronic interlocks in cruise missiles, the misfire was always an accidental launch rather than a technical malfunction.

> Cruise missiles have several layers of mechanical and electronic interlocks that preclude accidental launches. The launch event cannot be activated by a single person – SOPs simply preclude that. The BrahMos did not misfire; it was accidentally fired. That means it cleared the entire sequence of interlocks and safeguards, involving more than one operator....If "deviation from standard operating procedures" was the cause, why did the

> initial report mention technical malfunction? Or, are the two interconnected? As per sources, the accidental firing happened during an operational readiness inspection. Every squadron, unit and department in the station is on "high alert" during such inspection. Such inspections are meant to test the unit's readiness for action and detect gaps in the safety management of critical systems. In this case, it seems to have had the exact opposite outcome – critical failure of a system, designed with multiple interlocks, under the highest level of scrutiny. It doesn't get any worse than this.[48]

Kumar also brought up the important issue of whether the IAF could continue to assign "second-rung officers" to missile units and highlighted the strategically significant issue of the prevalent hierarchy within the IAF and its implications for regional security.

> One also hopes that the dreaded 'HR' angle doesn't find a place in this debacle. In the hierarchy of a fighter pilot-centric air force, missile units and their officers tend to sediment towards the bottom. The IAF's personnel branch would know how many grounded pilots (medical, competency or otherwise) eventually find their way into missile units. There may be some areas of overlap, but for the most part, these are different paradigms requiring different skill sets and, most importantly, attitude. With the focus shifting from manned to unmanned, manned-unmanned teaming, stand-off weapons, and nuclear-tipped missiles, the old playbook of assigning 'broken wings' and second-rung officers to such strategic forces merits a serious relook.[49]

Retired Air Commodore T.K. Chatterjee also pointed out the limited powers of the Court of Inquiry (COI) in a 29 August article and protested the harsh punishment meted out to the three IAF personnel found guilty.

> But what is more surprising is the alacrity with which the punitive measures were meted out by the government to the three defaulters, after a mere Court of Inquiry, which is only empowered to collect facts and apportion blame but has no powers to recommend punitive measures. There is no denying that the errors made by the missile crew in their respective operational and supervisory roles were grave since the incident could have had serious consequences internationally. But the AF (Air Force) Act provides for some more steps before a guilty is subjected to the coup-de-grâce....In my career of more than three decades, I have seen many acts of omission and commission committed. Have seen fully serviceable aircrafts getting written off in human error accidents, there have been instances of ships exploding or ramming into each other, fratricide supposedly in the fog of

war, but have not seen any instance of such hurried dispensation of justice....Hence such quickfire reaction of the government definitely points to some motive other than just exemplary retribution in order to correct a faulty system.[50]

Missile Launch as Deliberate—Part of More Assertive Indian Posture

Pakistani strategic and political affairs analyst Dr. Qamar Cheema hosted Indian Army veteran and Senior Consulting Editor (Strategic Affairs), Major Gaurav Arya of the right-wing Republic Media Network, on 27 August in which the latter confidently claimed that the BrahMos missile had multiple layers of safety and security and was handled by personnel who had graduated from India's top institutes including the Indian Institutes of Technologies (IIT's). Arya found it hard to believe that such intelligent operators would make such silly mistakes.[51] He also explained that the Pakistani demand for a joint probe was not possible as the Indian right-wing political leadership from former Prime Minister Vajpayee to Modi had been seared by the failure of goodwill outreach initiatives such as the Lahore bus diplomacy and invitation to the Pakistani ISI to jointly investigate the attack on the Pathankot air base.[52] This had given rise to a mindset of complete avoidance and circumvention of Pakistan.

Arya explained that the missile intrusion was part of a more assertive Indian posture wherein Pakistan's possession of nuclear weapons and the potential for escalation were not seen as major factors constraining Indian actions. He likened this to Pakistan's lack of concern over India's possession of nuclear weapons when it supported insurgency in Kashmir, Punjab, the 2001 Parliament attack, Akshardam temple attack of 2002 and the Mumbai attack of 2008. On being asked by Cheema as to whether there were rogue elements in the Indian military, Arya ended by reiterating his view that the India missile launch could not have occurred accidentally given the layered permission systems in place and was definitely a deliberate action undertaken by India.[53]

The Pakistani Strategic Community—Missile Launch Deliberate, HQ-9 System Acquisition

The intrusion into Pakistani airspace by the Indian missile led to sustained involvement of the Pakistani strategic community as it tried to ascertain the motives behind the missile's launch, the efficacy of Pakistani air defense systems, the delay in India's official statement, the response of the Pakistan government, the announcement by India of a Court of Inquiry, the results of the probe, the implications of the incident for regional security and other issues.

Speaking in a show on *ARY News* on 11 March, defense strategist Major General (retired) Ijaz Awan contended that the flying object was indeed a missile and did not have a warhead. He also claimed that the missile's violation of Pakistan's sovereignty and airspace was "not by default but by design." According to Awan, it was not credible that the guidance system of the missile was so erratic that it suddenly turned from its original trajectory which was heading southwards. It was a missile that was definitely launched by a missile unit. He opined that such a move could have many deep motives.

Firstly, India wanted to test Pakistani airspace observation, monitoring and management capabilities and wanted to determine the degree of Pakistani vigilance. Secondly, India wanted to determine Pakistan's capability to intercept a missile that could have had a warhead and that flew three minutes in Pakistani territory. Thirdly, India also wanted to ascertain Pakistan's diplomatic and political response.

Awan then explained that Pakistan's air operations systems consisted of three basic airspace management centers that were continuously monitoring the country's north, south and center. In case the missile had a warhead, the question of an appropriate Pakistani response would be an issue that will be discussed in the GHQ (General Headquarters) and in the Operations Directorate. He warned about Pakistan's lack of strategic depth with its center only about 450 km deep which Indian missiles can penetrate in its entirety and analyzed that the timing of the missile's launch seemed intended to worsen the prevailing domestic instability and political bickering in Pakistan.

Awan also connected the accidental launch with other lapses (6 kg of uranium being stolen in India and India's violation of Pakistani airspace during the Balakot air strikes in February 2019). He worried that any misunderstanding from a similar crisis in the future could have serious consequences and called on both India and Pakistan to discuss such issues while also enjoining the Pakistani government to involve the UN. Awan signed off by beseeching the Pakistani government to give a considered political and diplomatic response to the missile intrusion while keeping military preparedness high as the "fundamentalist" regime in India was capable of similar future transgressions.[54]

In *The Qamar Cheema Show* on 12 March, veteran strategic and political analyst Qamar Cheema questioned whether Pakistan too, like India, needed to acquire the Russian S-400 missile defense system, given the Indian missile intrusion. Cheema reminded his online audience that India has already acquired four batteries of the S-400 system from Russia with two batteries deployed on the Pakistan border, despite initial American threats of CAATSA (Countering America's Adversaries Through Sanctions Act) sanctions followed by eventual US acquiescence. He analyzed that while

Pakistan did not have hegemonic aspirations like India, it was determined to not live under an Indian-led order in South Asia with India as the policeman.

Cheema also explained that India's acquisition of the S-400 system and setting up of a multilayered defense shield as also a two-tiered ballistic missile defense system with high altitude interception capabilities, gave it both capability and confidence. To prevent future Indian missile transgressions, Cheema recommended that Pakistan complete the acquisition of the HQ-9, Chinese-made, Surface to Air Missile (SAM) system based on the S-300 with 200 km range and a flight ceiling of 40 km. He complained that negotiations were on since 2015 but had still not been completed. He also warned about the lack of political consensus and political stability in Pakistan which was preventing key political decisions on acquisition of defense hardware from being made. Cheema ended by recommending that Pakistan manufacture missile defense systems indigenously given the paucity of funding and inability to repair the frayed US-Pakistan defense relationship.[55]

Speaking to *SAMAA TV* on 11 March, Brigadier (retired) Nadir Mir, Chairman, Pakistan National Reform Movement, warned that India is preparing for war with Pakistan and that the preparation for this war is being done through the test firing of the missile into Pakistan. He also connected the missile intrusion to a recent claimed intrusion by an Indian submarine into Pakistani waters, to past Indian border intrusions into Pakistan and to terrorist attacks in Afghanistan and Baluchistan. He reckoned that these Indian actions were part of a sophisticated Indian war waged at multiple levels.

Mir recommended that Pakistan announce a policy for such kind of missile intrusions and questioned why Pakistani missile defenses were not used against the errant Indian missile, given that it could have landed in a Pakistani city causing casualties. He also analyzed that, even if the missile intrusion was an accidental Indian launch causing damage in Pakistan, the latter has to retaliate or else an emboldened India will fire missiles with warheads in the future. As an example, he brought up Pakistan's downing of an Indian fighter aircraft on 27 February 2019 and how this led India to stop further airstrikes after Balakot.[56]

Appearing on the same show, defense expert Amir Latif also agreed that the incident was not a mistake but a deliberate attempt by India to test Pakistan's nuclear deterrence capability. Explaining the technical features of the BrahMos missile including its ramjet engine and its capability of a low altitude trajectory, Latif explained that the missile travelled at an altitude of 34-35 km and then turned and fell near Mian Channu. Given this, Latif analyzed that the missile launch was an intentional move to test Pakistan's air defence system and an effort at strategic signaling by India that Pakistan should take seriously, signaling in return that it is fully prepared. Latif concluded by noting that the missile intrusion was part of provocative actions emanating from a fundamentalist and fanatic mindset of the regime across the border.[57]

Retired Pakistani Major General Inam Ul Haque in a 13 March article in *Tribune Magazine* attempted to discuss the "what, how and why" of the missile misfire as also its implications.

> First, the "what" side – the rear end of the "object" curiously matches the honeycomb structure at the tail of the Indian supersonic BrahMos cruise missile as per reports in the media and its photographs....A 2009 test failure was attributed to software problems, with missile missing the target by seven kilometres. So that is where the problem could be....A more versatile and lighter BrahMos-NG (Next Generation) is reported to have 290 km range cruising at Mach 3.5 speed in all army, navy and air force variants. It weighs only around 1.5 tons, is just 5 meters long and 50 cm in diameter. BrahMos-NG has lesser RCS (radar cross section), is hence harder to locate and engage by air defense systems. It was scheduled to test flight in 2022, and most probably that is the BrahMos that crashed in Pakistan....Moreover, given the Indian offensive posturing recently and the missile's significant features; this does not seem so innocently accidental. It could have been a deliberate effort to check Pakistan's Air Defence, readiness both at the PAF (Pakistan Air Force) and the Army levels.[58]

Software Glitch, Incorrect Entry of Coordinates, Incompetence of Missile Unit Personnel

In another piece on 17 March in *The Express Tribune,* Inam Ul Haque considered the possibility of a software glitch, incorrect feeding of coordinates and incompetence of missile unit personnel as the possible reasons for the missile launch. Pointing to the increasing technical sophistication of the BrahMos missile since its previous failed test in 2009 and connecting it with the rise of hardline nationalism in India, Haq concluded that the missile launch was no accident. He analyzed that it was in fact a deliberate move by India to test Pakistan's air defenses and response preparedness near the PAF base of Shorkot.

> One is amazed to see Pakistan's meek response, influenced ostensibly by *aman ki asha*; asking questions from India, that technically arise 'after accepting' the Indian narrative of 'accidental' launch....Missiles cannot be accidentally fired as the warhead, missile body and launch vehicle are stored safely in different places; and are prepared (mated) before launching. The onboard computer is, likewise, fed a predetermined target/data, a predesignated path/trajectory and pre-selected way points. And the

> missile is loaded/fired through a 'deliberate process' involving checks and counter-checks.
>
> Software glitches can lead a missile astray, like the 2009 case of BrahMos missing the target, but that was early on, and India's improvement of Inertial Navigation System (INS) involving US, Indian and Russian satellites, rule out this possibility. The other possibility is incorrect feeding of a 'way point', as the missile swerved mid-flight towards Pakistan, but repeated checks before launch should rule out this possibility either. Or else, it validates Indian ineptness and sheer incompetence, being unable to keep its sensitive programme in "capable and safe" hands....Add to this the narrow nationalism of war-mongering Indian media, and India's increasing intolerance towards Muslims/other minorities, the mix is toxic. So, India's offensive posturing, and BrahMos NG's significant features substantiate, it was a premeditated, technically prepared launch with predetermined and computer programmed trajectory, including a preselected waypoint to undertake the required maneuver in the direction of PAF Base Shorkot, to check/test Pakistan's air defence alertness and response preparedness.[59]

Expressing his views on Indian journalist Burkha Dutt's *Mojo Story* online channel on 16 March, Pakistani strategic affairs expert and journalist Mosharraf Zaidi expressed relief that the missile landed in Mian Channu and not in a densely populated area, causing casualties. Zaidi saw Pakistan's restrained response to the missile intrusion as a continuation of the de-escalatory responses it had adopted towards India since the Balakot airstrikes of February 2019. Urging that the incident should lead to introspective discussions in both countries, he informed that questions about the efficacy of Pakistani air defense systems and the need for proportionate retaliatory options were being discussed in Pakistan.

Zaidi also warned that "peaceniks" in Pakistan would now be under pressure to explain Pakistan's restrained reactions to such perceived provocations. Assessing that India does not have the stability or continuity in policy towards Pakistan since Modi's ascent in 2014, Zaidi revealed that he himself questioned Pakistan's capability and whether it could defend itself in case of a similar future contingency despite not being a national security hardliner.[60]

Missile Launch an Attempt to Normalize Limited Strikes, Self-Destruct Option Not Activated

Faizan Fakhar, Assistant Research Associate at the Islamabad Policy Research Institute (IPRI),[61] considered both the 'technical malfunction' and 'deliberate design' causative explanations for the missile launch in a 23

March article in *The Express Tribune*. He analyzed that the missile launch could have been a part of India's strategy to normalize limited strikes with narrow objectives against Pakistan and to test the latter's full-spectrum deterrence.

> There could be two possible explanations related to this event: technical malfunction or deliberate design....Even if it was a technical malfunction, it raises many questions on the Indian competence to handle high-end weapons, which are nuclear capable....In the nuclearised environment of South Asia, an accidental firing of nuclear capable BrahMos is highly risky since it is technically difficult for any available detection system to ascertain whether the incoming missile is armed with a payload or not. Moreover, with such high-speed missiles, it is highly challenging to intercept such projectiles in air....India has been seeking to "normalize" limited "surgical" strikes with Pakistan...The claimed Indian surgical strike (2016) and post-Pulwama attack on Pakistan (2019) also exemplified the aggressive posture of India....On 18th February, 2022, Yogi Adityanath, the current CM of Uttar Pradesh, explicitly stated, "BrahMos hub Lucknow will smile in the next surgical strike", indicating that the supersonic BrahMos will be weapon of choice for a surgical strike against Pakistan....If this was indeed a deliberate Indian design, the motivations could be multi-faceted, such as: to use it as a test case for a future limited conventional strike with narrow objective through supersonic missiles; to test India's own capabilities for an eventual first strike against Pakistan; and to test the threshold (of) Pakistan's "full spectrum" deterrence.[62]

Retired Air Marshal Shahid Alvi, Deputy President of the Lahore-based Centre for Aerospace & Security Studies, castigated the "international community, especially the United States" for its muted acceptance of the Indian explanation for the missile accident in a 23 March article for *PakistanPolitico*. He also pointed out that one civilian airplane on the Indian side and four on the Pakistani side could have been hit by the errant Indian missile and that its trajectory was only 20 km away from a "busy air corridor" where multiple international civilian airliners were travelling at the same altitude. Alvi also contended that, given the multiple mechanical and SOP-related fail-safes before any missile launch and the usual undertaking of maintenance operations by India in daytime (and not in the evening when the missile was launched), India's explanation regarding an accidental fire could not be taken at face value.

> Considering the range of BrahMos, it was all too logical that it would initially head elsewhere after being fired from Indian Air Force base Sirsa

> and having consumed some of its excess mileage (deception profile) would turn towards Pakistan Air Force (PAF) southern Punjab's base. It exactly followed its genetic coding, however, fell short, just like it did during the Balakot (wrongly fed coordinate) famed Indian strike.... Let's start with the dud explanation of "accidental" firing. The ill-fated missile was fired at 7 pm on Wednesday. In the Indian Air Force context, missile maintenance/inspections is not done in the evening; it is a daytime affair and preferably done on the maintenance day. Therefore, it was not a routine activity....Second, there are multiple fail-safes and in-built fail-safes in such lethal guided systems both in terms of physical and soft prohibitions which need to be removed prior to firing a live missile. These sets of prohibitions are deliberately incorporated by OEM to avoid such accidental occurrences during maintenance and training activities. Third, there are also specific sets of operational procedures and SOPs which govern the process of firing a live missile. These procedures are designed in such a way that a missile is fired only on/towards an intended target when deemed necessary by the user. Moreover, if the system was under maintenance as claimed, then as per the SOPs, dummy missiles are mounted in place of live missiles prior to maintenance, Furthermore, in order to bring a system back into active operational state, various checks and procedures need to be performed which are documented in black and white. Generally, maintenance of such guided weapon systems is carried out in storage and even if done in open then launch direction is kept towards a safe firing range and away from vulnerable areas. And, last but not least, how did the "self-destruct" sequence magically vanish even if the missile was launched by mistake? Given these logical and operational reasons, one simply cannot swallow India's ludicrous claims.[63]

Jinnah Institute, a think-tank with offices in Islamabad and Karachi, solicited views from experts to analyze the missile incident and its implications for "South Asia's crisis stability" which were consolidated and published on its website on 25 March. Senator Sherry Rahman, the Institute's President called for a new level of Pakistani preparedness for an increasingly dangerous regional security environment created by Indian assertiveness and defects in its SOPs for handling missiles.

> The latest episode also tells us two things even if Indian claims are accepted: that Indian missiles are sitting trained on Pakistani territory, and that "slippages" during routine maintenance may well claim lives. It hasn't spiralled this time, but incidents like this are exactly the stuff of nuclear safety simulations where incompetence leads to escalation....A country that is ranked 3rd highest in military spending ($72.9 billion per year) just after the US and China, and fond of indulging global power

> delusions, had no compunctions about adopting a cavalier public posture about the scale and nature of the harm it is committing. We recognize the new precedents, and know full well that no country would have gotten away with launching an 'accidental' missile anywhere....Missile 'accidents' like this convince all actors in a nuclear environment that the future is unpredictable and lethal....Pakistan too needs to consider another level of preparedness now.[64]

Dr. Rabia Akhtar, the Director of the Centre for Security, Strategy and Policy Research at the University of Lahore and a member of the Prime Minister's Advisory Council on Foreign Affairs, suspected that the missile intrusion may have been part of a deliberate Indian strategy, disguised as an accident, to collect data on Pakistani air defenses during peacetime without eliciting a Pakistani response. She also questioned the efficacy of the Pakistani air defense system, given its unwillingness/inability to shoot down the Indian missile.

> It (missile accident) adds another layer of instability to the already fragile crisis dynamic in South Asia....While the Indian official statements refer to this as an accident, Pakistan has reasons to speculate further and wonder if India was testing gaps in Pakistan's air defense. It is a legitimate question given that India can only do so during peacetime based on the data it obtains by probing Pakistan's air defense preparedness. And a missile strike disguised as an accidental launch landing in the enemy territory without it being destroyed by the PAF is a very good data point. While Pakistan's response was that of restraint as is being lauded, allowing an 'incoming object' to travel hundreds of kilometers into Pakistan's airspace without engaging with it, does not lend much confidence upfront about air defense preparedness.[65]

Sitara Noor, Senior Researcher at the Islamabad-based Centre for Aerospace and Security Studies, critiqued India's failure to initiate the self-destruct mechanism once it realized that the missile had been accidentally launched. Recommending that India and Pakistan expand the 2004 hotline agreement to include pre-notification for cruise missiles, Noor warned Indian strategists to not mistake Pakistani restraint for lack of capability.

> Accidents may happen as no system can claim to be completely failsafe, but as a responsible nuclear weapon state, India should have taken certain immediate corrective measures such as initiating the self-destruct mechanism during flight and informing Pakistan to avoid an inadvertent crisis....The non-response makes one wonder whether this incident resulted from a safety failure or a security breach? In either case, it has

exposed India's poor strategic culture and inefficient response mechanism that could pose serious ramifications for regional peace...India is steadily increasing its alert level through various means such as canisterisationof missiles, diluting its No First Use (NFU) pledge, and gradually shifting towards a counterforce posture. A mishap of this nature in such a volatile environment could have very easily triggered a serious crisis....India and Pakistan must strengthen the 2004 Hotline agreement—that should have been used as a means of engagement in this case. The two sides may also consider expanding the agreement to include pre-notifications of both, flight testing of ballistic missiles and of cruise missile tests as well... Lastly, Pakistan's measured response that helped avoid a crisis should not interpreted as a sign of weakness by India. Many Indian analysts are incorrectly suggesting that the incoming missile took Islamabad by surprise and that it did not increase its alert level. Drawing such inaccurate conclusions may lead to faulty calculations and decision making in the future with grave consequences for the two countries and the region.[66]

PAF's Decision to Not Intercept the Missile, Pakistan Must Reverse-Engineer It

Dr. Rabia Akhtar also elaborated on her aforementioned views in a 31 March online panel discussion titled "Skirting the Nuclear Threshold: Conflict De-escalation and Strategic Stability in South Asia," organized by the New Delhi-based Council for Strategic and Defence Research. She expressed dissatisfaction over the PAF's operational conduct during the missile accident. She questioned the PAF's assumption that the incoming missile did not have a warhead (despite it coming from the territory of a "declared enemy") and its resulting decision to accord the missile a "red carpet treatment" by tracking it for seven minutesfrom its launch in India to its impact in Mian Channu. Even if the missile has a small radar cross section and was travelling at high speed, an attempt should have been made to engage and intercept it. After all, Pakistan did not possess enough information about India's peacetime experimentation and its intentions. Akhtar speculated that the PAF might have not engaged with the missile as its flight path was through a busy air corridor with Saudi and Qatari commercial airlines in the vicinity.

Terming the PAF's assumption of the incoming missile as a dud as a dangerous one even during peacetime, Dr. Akhtar termed its decision to not initiate tactical actions as a good learning opportunity which should stimulate "alternative thinking" in the Pakistani strategic community. She also brought up the severe trust deficit between India and Pakistan due to lack of dialogue and the resulting inclination in South Asia to resort to conspiracy theories. While terming it possible to accept the Indian explanation that the missile launch was unintentional, she also expressed an inclination as a scholar of regional security to probe whether the missile launch was an unintentional act or disguised as unintentional. Akhtar questioned the inordinate Indian delay of 48 hours in informing Pakistan about the accidental launch and criticized it as absolutely unacceptable.

Questioning India's decision to commission a unilateral Court of Inquiry and the reliability of its findings for Pakistan, Akhtar expressed doubts over the honest sharing of the inquiry's findings with the Pakistani leadership. Expressing worries about the possibility of similar accidents in the future due to human or technical error (including an accidental Indian drone launch) in a manner akin to the numerous Cold War-era accidents, Akhtar called on both countries to engage and share their best practices concerning protocols and SOPs. She also framed the Indian risk perception of a two-front threat from China and Pakistan as an "Indian narrative" which India uses to express wariness about being a party to any bilateral strategic restraint dialogue. Identifying the 1991 agreement on air space violations as the right instrument for reducing risks, Akhtar called for the expansion of its scope to include armed drones and accidental missile launches. Making a case for Pakistan to reverse engineer the missile, she averred that it would be "stupid" on Pakistan's part to not undertake such an effort.

Indian Missile Developers' Morale, Command and Control Reliability, Fog of Disinformation

Dr. Ahmed Saeed Minhas, a Karachi-based security expert and pro-Vice Chancellor of DHA Suffa University reasoned in a 27th March article that the Indian attempt to frame the missile launch as a low-level incident was an attempt to prevent morale deterioration in the institutions that were part of the BrahMos manufacturing and deployment ecosystem.

> The Indian deliberate effort to keep the incident low profile probably meant four main purposes. One, India wanted to avoid demoralisation of its strategic weapons' development organisations, mainly the Defence Research and Development Organisation (DRDO), the Indian Ordnance Factories Board, BrahMos Aerospace Private Limited and Bharat Electronics Limited (BEL), which are actively involved in production of missiles in India. Secondly, India may have feared that the planned bulk production of BrahMos missiles could be severely affected for the purpose of exporting it to other countries. It may be noted that in January 2022, Philippines' Defence Ministry signed an export contract of undisclosed numbers of BrahMos missiles worth $374 Million with BrahMos Aerospace Private Limited...Third, the Indian strategic community does not want to burden itself with increased criticism of its strategic command and control both at domestic and international levels. Last but not the least; the accident of BrahMos missile can question the credibility of

> Indian research efforts to develop BrahMos-II, having speeds of Mach 7-8 and an enhanced range up to 1,000 km with an aim to boost striking capability in shorter time frame without being detected using cruise technology. It is likely that the accident involved the advanced version of BrahMos, which is currently in research and development phase.[67]

In a 29 March article, Haris Bilal Malik, a researcher at the Centre for Aerospace & Security Studies (CASS), Islamabad, raised doubts over the official Indian explanation of an accidental launch and worried about its implications for the reliability of India's nuclear command and control system.

> The Indian claim of 'accidental launch' is suspicious and independently unverifiable. Cruise missiles, regardless of their land, sea, and air-launched versions, are guided missiles that follow a specific flight trajectory. It is difficult to accept the Indian explanation that the cruise missile was accidentally launched and entered Pakistan's air space on its own.
>
> Furthermore, even if it was an accidental or unauthorized launch, it raises serious questions about the credibility of the Indian nuclear command and control system, operational protocols of its strategic delivery systems in general and the safety of the Indian missile arsenal in particular. These safety and security issues pose grave risks to regional stability because India has been pursuing extensive and all-encompassing military modernisation, including the development of nuclear-capable supersonic cruise missiles....incidents such as these indicate that the Hindu extremist government currently ruling India and its unprofessional military cannot be trusted with such lethal missile technology....It is equally important to further enhance our air defence system, particularly against the growing threat of India's supersonic cruise missiles.[68]

Dr. Asma Shakir Khawaja, Executive Director at the Center for International Strategic Studies, AJ&K (Azad Jammu and Kashmir), criticized the delays and ambiguities in the Indian government's statements in a 30th March article in the *Global Village Space*. Suspecting that India's "Hindutva radicals" were trying to establish a new norm of missile intrusions across the border in a manner akin to limited surgical strikes, Khawaja recommended that India and Pakistan incorporate pre-notification requirement for missile intrusions under the Preventing Air Space Violations Agreement of 1991 which focuses mainly on violations by combat aircraft.

> Here another question arises is India 'deliberately' trying to establish a new norm of accidental fire in South Asia? It is noteworthy that in a volatile and nuclear region of South Asia, such a norm can neither be

established nor it can bring any good to the region. Such an act will only make the situation of peace and security more vulnerable to the extremist Hindutva mindset, especially when it is still to be established if it was a mistake or the Hindutva radicals have access to such highly sophisticated weapon systems....On 11th March, in an official statement, the Indian government stated that "in routine maintenance, a technical malfunction led to the accidental firing" of the supersonic BrahMos missile. However, on 15th March, the Indian defense minister declared in the Indian parliament that this accidental firing occurred during "Routine Maintenance and Inspection." Furthermore, few journalists wrote that this accidental firing was caused during the "Simulation Exercise." These three different explanations only contributed to creating the fog of disinformation...The fog of disinformation and inappropriate/misleading strategic communication by Indian leadership have multiplied the nuclear risks in South Asia.[69]

Need for a Joint Investigation—Indian Launch Authorizing Authority and Civilian Primacy

The Pakistani strategic community also came together on common forums such as the 1 April webinar titled "Indian Irresponsible Strategic Behaviour: Intentions, Regional and Global Consequences," organized by Rabita Forum International, a Karachi-based organization. The aim of the webinar was a detailed and multifaceted discussion of the missile launch. Experts comprising the panel were Ambassador Munir Akram, permanent representative of Pakistan to the United Nations, Former Ambassador to the UN Zamir Khan and Dr. Syed Muhammad Ali—Director, Centre for Aerospace and Security Studies (CASS).

The Chairman of the Rabita International Forum Mr. Nusrat Ali Mirza began by summarizing the events leading to the accident and by wondering how the BrahMos missile, which India had supposedly deployed in February 2019 near Rajasthan to threaten the Pakistani city of Bhawalpur and the Rafiqi Air Base had suddenly been accidentally launched into Pakistan? Ambassador Munir Akram urged Pakistan to sensitize the international community regarding India's reckless and dangerous behavior. He also reiterated the eight questions posed by the Pakistani Foreign Office to India (covered in detail in Chapter 2) and called for a joint investigation. Akram called for a broader discussion of the implications of the missile misfire for civil aviation at relevant organizations such as the International Civilian Aircraft Organization (ICAO). He also informed that Pakistan was contemplating a discussion of the accident with the Missile Technology Control Regime (MTCR) to which India is a party while Pakistan is not.

Ambassador Zamir Akram expressed skepticism over India's initial "technical malfunction" explanation and asked why India did not use the Director General of Military Operations (DGMO) hotline to inform Pakistan and why its acceptance of the accident only came 24 hours after the Pakistani DG-ISPR's press conference. He also worried about the security ramifications of the accident given the missile flight time (and the resulting alert time for the country's airspace it intrudes) being less than three minutes between India and Pakistan. Akram also drew attention to the Indian submarine intrusion of Pakistani waters and enjoined Pakistan to determine whether the missile misfire was an accident or a deliberate act.

Dr. Syed Muhammad Ali of CASS reckoned that Pakistan was right in rejecting initial explanations by the Indian government as a cruise missile like the BrahMos requires an elaborate launch protocol of between eight to ten steps and pre-fed data to fly along a pre-determined path. A technical malfunction or an accidental launch would require errors across the entire series of steps. The cruise missile cannot accidentally launch at the press of a button and cannot technically change direction as abruptly without operator control. He also questioned whether the missile had an in-built self-destruct system and why the missile operators did not activate it when they realized that the missile had veered off course. Dr Ali also warned that, the accidental launch pointed to India's cruise missiles being in launch-ready state which raised serious questions regarding authorizing authority (Prime Minister, Strategic Forces Command, local missile unit commander) and broader adherence to the precepts of civilian primacy in launch-related decision-making and doctrinal premise of No First Use (NFU).[70]

Cruise Missiles under 2005 Agreement, Internal Opposition from BrahMos Developers

GulRukh Nafees, a research associate at the Centre for Law and Security in Islamabad called on India and Pakistan to conclude an agreement with clear rules to manage similar future contingencies in a 11 April article for *Geotv*.

> Firstly, it (Agreement on Pre-notification of Flight Testing of Ballistic Missiles) applies to ballistic missiles and not cruise missiles. Secondly, it does not provide for an accidental launch. Nonetheless, there is a hotline between the director-general military relations of Pakistan and India, but what's surprising is that even this option was not considered during the accidental misfire....In fact, the whole incident was acknowledged two days after its occurrence...both the countries should draft an agreement explaining the conduct and strategies that shall be adopted in case such an incident takes place again.[71]

Natasha Khan, a Research Assistant at the Islamabad-based, Centre for Strategic and Contemporary Research,[72] criticized the ambiguities in India's No-First Use (NFU) policy. Khan, like other Pakistani analysts, called for cruise missiles to be brought under the ambit of the 2005 Agreement on Pre-notification of Ballistic Missile Tests.

> The BJP's aggressive bearing and India's constant ambiguity over its nuclear Non-First Use (NFU) policy fall within the purview of such conduct. India has deliberately maintained this ambiguity, often citing the threats from China and later Pakistan as the reasons for maintaining Credible Minimum Deterrence (CMD), a clause that allows it to respond to threats – real or assumed – as needed. There is an alarming likelihood that the command and control authority's negligent control of weapons caused the unauthorised or accidental launch of the missile....This (2005) agreement can be extended to cruise missiles to minimise inadvertent nuclear risks.[73]

Muhammad Ali Baig, a research associate at the Institute of Strategic Studies Islamabad (ISSI), analyzed in a 19 April article that the Pakistan Air Force (PAF) did not use its surface to air missiles to intercept the Indian missile and could have used its electronic warfare capabilities to jam it, resulting in its crash landing at Mian Channu. He also brought up suspicions of an intentional strike by an India with its missile policy taken over by Hindu extremists such as incumbent Chief Minister Yogi Adityanath whose Uttar Pradesh state hosted a BrahMos manufacturing facility and who was recently victorious in the state election.

> It is evident from DG-ISPR's Press Briefing that the missile was detected well inside the Indian territory; however, it is worthy to probe whether any attempt was made by Pakistan to intercept the incoming flying object. Apparently, the data gathered from radar and the velocity of the flying object made the operational decision-makers to conclude that it could have been a missile and not an aircraft. Ostensibly, no surface-to-air missile (SAM) was fired to intercept the incoming missile; however, the PAF could have possibly used its electronic warfare capabilities to confuse or even jam the guidance and navigation systems of BrahMos resulting in crashing the missile after traveling 124 kilometers inside Pakistani territory. It is prayed that the international community especially the great powers take heed of India's irresponsible state behaviour and keep a close watch of its advanced military arsenal from falling into the hands of religious fanatics such as the hate peddler Uttar Pradesh Chief Minister Yogi Adityanath who openly advocated the use of BrahMos cruise missile against Pakistan and its citizens...If India continues to run from initiating

a joint investigation, then the hypothesis of an intentional and premeditated strike against Pakistan could gain traction.[74]

In another article a month later, Baig also speculated about internal opposition from the developers of the BrahMos leading to the reason for the incident being changed by the Indian government from the initial "technical malfunction" in the Defence Ministry's statement to human error in the preliminary results of the Indian Court of Inquiry. He also criticized the Indian missile intrusion as an unauthorized launch with "no legal consent by the political authority."[75]

Indian Insider Threat, Shift in Nuclear Doctrine, Pakistan's "Quid Pro Quo Plus" Strategy

The Center for International Strategic Studies (CISS), an Islamabad-based think-tank, conducted a round table discussion on "Indian Missile Fiasco: Technical Malfunction and Failure of Diplomacy" on 21 April with Dr. Syed Muhammad Ali, Dr. Zafar Jaspal, Dr. Rizwana Abbasi and Ambassador Ali Sarvar Naqvi.

Dr. Ali, the Director of the Centre for Aerospace and Security Studies (CASS) organized his remarks along three aspects: safety, security and intentional use. He was categorical that India should have used the hotline to warn Pakistan that the missile launch was not intentional. He also raised several questions at the operational level such as whether the Indian forces have a two-man rule for authorizing missile launch and whether they possessed permissive action links? He also questioned whether the inspection exercise allows the missiles to be kept at high readiness levels and whether it was possible for an Indian missile regiment commander to launch the missile and to override some procedures and technical requirements in a classic manifestation of the "mad colonel" dilemma.

Asking whether the launch was indeed intentional, Ali analyzed that the incident was indicative of the risk from insider threat. He further speculated that the incident could have resulted from ideological motivations of extremist insiders, due to psychological instability, or due to a lack of trust between political and military authorities. He also questioned the authorization protocol and procedure for launching missiles of a strategic nature in India and whether missile launches required top-level authorization or could be launched by a missile commander of their own volition. He also worried that the latter possibility would contradict India's doctrinal commitment to a "centralized, assertive nuclear control." He also brought up the "worrying possibility" that Indian missile forces (commanders) were not under the direct control of the political leadership as per India's doctrinal commitment, especially in light of a Group Captain being found culpable by India's COI. He reiterated that the mad colonel dilemma, the insider threat resulting from it and the potential of a deliberate launch threatened deterrence stability and crisis stability much more than an accidental launch.

Ali pointedly wondered whether or not the Indian missile system had a "fail-safe, fail-deadly positive and negative controls" built into its forces? Focusing on the doctrinal aspect, Ali explained that the accidental launch reflected a dichotomy "between the political commitments made by India under its declaratory doctrine and the actual developments on ground in terms of readiness level," force posture and operational procedures. He was clear that the incident raised serious questions about the safety, security and reliability of India's strategic forces and that both technical and human causes deserved scrutiny as either of them could have caused the accident.

Ali further speculated that India is perhaps operationalizing a counterforce posture and cited nine reasons to support his claim: 1) readiness levels are high, 2) missile forces are canisterized, 3) missile forces are kept in a ready to fire state, 4) there is greater development of cruise missiles rather than ballistic missiles with a greater emphasis on accuracy rather than scale of destruction, 5) launchers are not fixed in silos but are mobile, 6) warheads are not being separated from delivery systems, 7) there is high investment in I2SR capabilities which is inconsistent with the no-first use doctrine, 8) the continuing increase in the size of the arsenal and 9) the development of BMD capabilities.

Ali also analyzed that India's nuclear strategy had shifted from an emphasis on massive retaliation to an emphasis on flexible response. He also did not rule out the possibility of a deliberate launch by India with later statements intended to dissuade Pakistani retaliation. He also contended that notwithstanding Indian explanations, the Pakistani response would have been different had the missile landed in a more populated area and caused casualties. Ali also recommended that Pakistan continue with its "quid pro quo plus" policy to Indian provocations but also calibrate its response based on the motive and nature of a future incident and whether it was unintentional or deliberate in order for the response to be proportionate.

Ali concluded his detailed analysis that the accidental launch was a grave incident and an "unprecedented one between two nuclear armed states" and that the international community must pay greater attention to the safety and security of India's nuclear and strategic weapons systems. Highlighting the three recent tests of the BrahMos missile conducted after the accidental launch by India, Ali surmised that India seemed to be addressing certain operational gaps that were discovered during the accident. He ended by informing that Pakistan's test of the Shaheen-3 missile a month after the accident could be interpreted as a signal to warn India that its entire territory including Andaman and Nicobar Islands could be vulnerable to Pakistan's strategic capabilities.

"Dual-Use" BrahMos Missile, DRDO Nervousness, Weakening of Indian NFU Doctrine

Speaking next, Dr Zafar Jaspal, Professor at the School of Politics and International Relations, Quaid-I-Azam University, Islamabad, analyzed that India had been working on both hypersonic and supersonic missiles and had been receiving technological assistance in this from other countries. He questioned whether these countries were taking into account the safety and security-related concerns surrounding the Indian nuclear and missile programs. Jaspal reckoned that there were three possible causes for the accident: a) accidental, b) authorized and c) inadvertent. He analyzed that, even if the Indian Prime Minister's Office (PMO) and the Ministry of Defence (MOD) pre-authorized this test and the Indian Air Force (IAF) then scheduled it on 9th March, the nature of this accident still qualified as an unauthorized missile launch.

Jaspal reminded that the BrahMos is one of the fastest supersonic cruise missiles in the world and both the Indian Army and the Air Force had been using it. He also lamented that had the missile hit a commercial aircraft leading to passenger deaths, the West would still have concluded that the projectile was a Pakistani rocket launched by a Pakistani non-state actor. If the missile had crashed in a populated area causing damage to life and property, Pakistan would have been forced to respond because the Pakistani public would not have accepted that it was just an accident. In fact, Pakistan's "quid pro quo plus" strategy demands such retaliation.

A Pakistani military response could have triggered a conflagration because of misunderstanding and miscommunication. Jaspal also worried that a technological error in the errant missile's navigation could result in India's accidental use of nuclear weapons as the BrahMos missile is dual use. He lamented that the incident had major security implications for South Asia but was not paid attention to by the West as the European nuclear flashpoint (Russia-NATO) has now replaced the South Asian nuclear flashpoint. He also accused the West of not having the courage to question India given its designation of the latter as an ally to counter China. He surmised that the overall response of the international community had been irresponsible.

Jaspal also summed up India's response to the incident as one of "nervousness" either at the level of the Defense Research and Development Organisation (DRDO) or at the "command level" due to the push and pull within the Indian armed forces wherein one faction supports a transition to "integrated theaterisation" and the other wants to retain the existing system of Corps. He further explained that one possible reason for the nervousness was the lack of clarity over who is the "commander of the strategic forces."

He also surmised that the DRDO had failed to achieve proficiency in missile technology despite decades of research and that it was getting assistance from the Russians which it was reverse engineering. India was also getting assistance from the Israelis but the DRDO was not capable enough to absorb their assistance. This combined with lack of effective security protocols and technical safeguards raises the risk of unexpected consequences.

Connecting the missile intrusion to the claimed violation of Pakistani waters by an Indian submarine a few days prior, Jaspal raised the possibility that the latter was an attempt to measure Pakistan's "naval preparedness" and the former could be understood to be the IAF's attempt to measure Pakistan's "defence preparedness." He alleged that the IAF had been nursing an "injured ego" since the Balakot air strikes when one of its pilots, Abhinandan Varthaman had been shot down, captured and later released by the PAF. Analyzing the implications of the missile incident, Jaspal contended that the launch had exposed "grave weaknesses" in Indian technology and safety systems revealing two things in particular: 1) The BrahMos missile is not technically advanced enough to be effectively deployed and 2) the Indian military is incapable of operating this missile system.

Commenting at the doctrinal level, Jaspal contended that Indian strategic pundits are swapping no first use for first use option in its nuclear doctrine (and transitioning from non-deployment to deployment of nuclear assets at sea). This was clearly announced by the Indian Defence Minister in 2019 about the quashing of the NFU concept in their military doctrine. Jaspal situated the aforementioned developments as part of efforts by the Indian strategic community since 2011 to get rid of the No First Use policy with a faction supporting its retention, a second faction calling for its removal and a third recommending an "operationally first use—rhetorically no first use" ambiguous policy.

Informing that even the US does not possess the capability of intercepting a supersonic missile, he recommended that the only "defence is deterrence" and the ability to demonstrate that Pakistan has a retaliatory capability. Jaspal understood the recent Shaheen-3 test by Pakistan as a signal to India that "we are in a position to bleed you." He also averred that missile technology in general cannot be entirely trusted and technological errors can occur necessitating the requirement for hotlines to be "primed and kept on the ready" during a missile test. He further recommended that cruise missiles be brought under the pre-notification requirement of the 2005 agreement.

India's Violation of Accords, Reactivation of Hotlines and Indian Nuclear Safety Lapses

Rizwana Abbasi, Associate Professor in the Department of International Relations at the National Defence University, Islamabad and the third speaker at the aforementioned round table discussion, contended that India had violated the India-Pakistan bilateral accords of 1991, 2005 and 2007 and had undermined global aviation rules by allowing the accidental launch. She also pointed out that India was bound to give immediate notice of nuclear accidents under the 2007 agreement. She questioned the Indian delay in informing Pakistan that the missile launch was an accident and averred that the incident violated India's obligations under Article 2, paragraph 4 of the UN Charter. Abbasi analyzed that India's not using the military (DGMO) and the Foreign Secretary-level hotlines suggested its "irrational nuclear behavior" and its hubris resulting in failure to engage with Pakistan and the resulting failure of diplomacy. She contrasted this with continuing US-Russia communications, even during the Ukraine crisis. Like Jaspal, Abbasi also viewed the missile intrusion in the context of the interception of an Indian submarine probing Pakistani waters and the Indian Army Chief's warning to Pakistan of pre-emptive strikes. She recommended that a joint probe was required to accurately establish the facts surrounding the incident as it had violated Pakistan's airspace, the spirit of bilateral agreements and international aviation protocols.

She also supported the joint probe as crucial to alleviate misperceptions, build trust and "mitigate future risk" and cautioned that a "one-sided" Indian probe would not satisfy Pakistan. Abbasi explained that the incident had pushed India and Pakistan towards greater nuclear weapon readiness and this new situation needs to be deftly managed. A votary of risk reduction measures, Abbasi reckoned that the most urgent requirement was for the reactivation of "strategic communication channels" and the addition of new ones for "crisis management." She also suggested that the 2005 agreement needs to be formally expanded to include cruise missiles including air launched and sea-based missiles.

Abbasi re-emphasized the need for India and Pakistan to communicate using a "secure, high-level hotline" in case of a similar accident in the future and beseeched both India and Pakistan to agree to not test missiles near the India-Pakistan border. She underscored the need for restraint measures to be taken against the deployment of destabilizing systems that could affect arms control and strategic stability. Abbasi also highlighted the need for dialogue between India and Pakistan to clarify the missiles that are assigned conventional roles and the ones assigned "strategic roles." The DGMO hotlines needed to be kept open, despite ongoing mistrust. According to Abbasi, the 2007 agreement should be amended to explicitly mention delivery systems and to cover "inadvertent accidental firings of missiles." Urging India and Pakistan to minimize the ongoing trust deficit, she recommended a reactivation of a political dialogue on all issues including Kashmir. She also suggested that India and Pakistan introduce "code of conduct" dealing with rules for the media during a crisis as the media plays a sensational role in escalating and not de-escalating the crisis.

Abbasi surmised that the accident also highlighted the serious need to "clarify the alert status of missiles" and the targeting practices between the two countries. She warned that India would be gravely mistaken if it drew the wrong conclusion from the "misadventure" that Pakistan would be similarly restrained if a similar missile intrusion proven to be deliberate on part of India occurred in the future. Any nuclear weapons state is likely to perceive an intruding dual purpose missile as the highest of threats and take response measures it deems appropriate. She reminded that luck was also an important factor as the lack of collateral damage restrained Pakistan's reaction.

Abbasi also linked the missile accident to the broader safety and security lapses in the Indian nuclear complex, as evidenced in the claimed numerous unauthorized purchases of Indian nuclear material (uranium theft) since 2016 in the commercial and black market. Dispelling notions of Pakistan's domestic political crisis preventing it from responding appropriately to the missile intrusion, she reminded that a Prime Minister was appointed to succeed Imran Khan and key institutions such as the National Command Authority (NCA) and those under it continued to function. However, in case of future crises, such institutions would need to be strong, resilient and capable of absorbing shocks.

Abbasi ended with a critique of the Indian response to the accident which initially claimed technical malfunction as the cause and later pivoted to the "narrative of human error" after the results of the CoI. She alleged that the pivot was undertaken when it was realized that the technical malfunction angle would impact Indian exports of the missile.

Ambassador Ali Sarvar Naqvi brought the discussion to a close by observing that the IAEA was the best international agency to address the accident as they have a "proven record and full apparatus for handling safety-related matters" including a Department of Safeguards, a Department of Security and qualified experts. He analyzed that India was fast building a triad of BrahMos cruise missiles and is deploying these missiles as the "key capability" to execute its counterforce ambitions against Pakistan and China. He reckoned that the missile accident had exposed the "chink in the armour" of this counterforce strategy.[76]

Indian Missiles in Launch-Ready Mode, Pakistan's Decision to Not Shoot Down Missile

Sabina Babar, a Research Officer at the Islamabad-based Strategic Vision Institute (SVI), reckoned in a 22 April article that India's accidental launch of a supersonic missile into Pakistan should impel Pakistan to bolster its supersonic and hypersonic missile capabilities.

> Pakistan has recently tested its ballistic missile Shaheen III, which is a defensive addition to its weaponry, however in the supersonic and hypersonic domain Pakistan still lags behind India. In response to the recent

> attack, Pakistan needs to focus on boosting its capabilities in supersonic and hypersonic missile technologies. So far, Pakistan has viewed the hypersonic race in the Indo-Pacific as a power competition of ascendency between the USA and China, however with recent incidence, Islamabad is cautious about the security and safety against supersonic and hypersonic on its border. It is likely that China will assist Pakistan in developing a Hypersonic Glide Vehicle missile (HGV)....Pakistan has built modern subsonic cruising Babur/Harbah and Ra'ad series along with the CM-302 Type-054A/P frigate programme. It seems Pakistan will pursue its own supersonic-cruising LACM through potentially the HD-1A, which China marketed as an option for the JF-17.[77]

Dr. Rahat Iqbal, Associate Director at the Center for International Strategic Studies (CISS), Islamabad opined in an 18 May article that the Indian accidental misfire was a serious matter that had been ignored by international institutions and would be interpreted by Pakistan as Indian missiles always being in launch-ready mode, with serious implications for South Asian stability. She also disputed the accuracy of the claim made by a member of the Indian strategic community that there were different command and control mechanisms for nuclear and conventional weapons.

> Regrettably, the incident was not taken as a serious matter by the global community and nuclear-established regimes, including the United Nations Security Council (UNSC) 1540, International Atomic Energy Agency (IAEA), International Panel on Fissile Material (IPFM) and global security organisation such as Nuclear Threat Initiative (NTI)....The misfiring of the missile indicated to Pakistan that the Indian missiles are always on the ready mode to launch without an appropriate command and control system....An Indian military officer Lt Gen D.S. Hooda, while speaking during a webinar organised by Council for Strategic and Defense Research (CSDR), New Delhi, claimed that there are different command and control mechanisms for nuclear weapons other than conventional weapons. However, this seems to be a faulty claim. Differentiating between the conventional and non-conventional arsenal in the context of missiles is difficult to measure at the time of the fire, particularly the supersonic cruise missile.[78]

Samran Ali, a research officer at the Center for International Strategic Studies, Islamabad, explained the considerations behind Pakistan's decision to not retaliate to the errant Indian missile in a 20 May article. He also highlighted how Pakistan's layered defense system would still leave it vulnerable to Indian cruise missiles, and explained the various proportionate retaliatory options to deter an Indian missile attack.

> Pakistan Army had inducted in October 2021 the HQ-9/P High-to-Medium Air Defence System (HIMADS) from China which can engage cruise missiles along with aircraft. It is the first layer of Pakistan's multi-layered air defence network. It has a 100+km range against aircraft and other targets. However, the engagement range against targets with a low radar cross-section is maybe around 25km, according to Janes. However, it is not confirmed whether the system was operational when the 9 March incident occurred. Before acquiring the HQ-9/P, Pakistan was already operating HQ-16 medium-range air defence system, which can also target cruise missiles. Its missiles can hit targets between an altitude of 400 to 10,000 meters and at a range of 40km. These systems do equip Pakistan with the anti-cruise missile capability, but they still have deficiencies against cruise missiles like any other air defence system in the world. Furthermore, the deployment location of air defence systems also factors in intercepting the targets. This layered air defence network of Pakistan is deployed to protect key installations and high-value assets and may not be available to cover every inch of the country. The acquisition and maintenance of air defence systems also have huge financial implications that dictate their limited deployment.Upon finding it as a single object approaching, Pakistan thought it was sufficient that the missile did not pose any threat to a high-value target in the country and refrained from taking any action in the military domain. For a retaliatory strike in a similar manner, Pakistan has a variety of cruise missiles such as subsonic Babur and Ra'ad and ship-launched CM-302 supersonic anti-ship/anti-surface missile, which travels at Mach 3 speed for up to 500km.[79]

Missile Accident a Minor Incident, Indian Military's Cover-Up, Proportionate Responses

Dr. Muhammad Muhammadi and Dr. Rahat Iqbal, both Associate Directors of Research at the Islamabad-based Center for International Strategic Studies (CISS), categorized the Indian missile misfire as a minor incident in a 31 May article for the *Daily Times*. Given the unwillingness of both India and Pakistan to escalate conflict even during periods of heightened tension due to the possession by both of nuclear weapons, the authors contended that incidents such as the missile intrusion would not disrupt the stable nuclear deterrence in South Asia.

> In a broader spectrum, the possession of a nuclear arsenal is the constant factor in maintaining the equilibrium among the conflicting states, particularly between India and Pakistan. During the Kargil crisis of 1999, both India and Pakistan adopted the ladder of de-escalation from escalation, as both were the nuclear power states by that time. Since then, it's been almost 24 years that both the states did not venture into full-fledged

> or limited war....The development of several missiles such as Shaheen-III, Ababeel, Babur, Nasr and some others could be considered a tool of deterrence for the state. Therefore, a stable nuclear deterrence is intact and does exist in South Asia. There is a possibility that scenarios such as the Balakot incident, revocation of articles 370 and 35A, illegal annexation of Kashmir and the BrahMos Missile incident may have negative implications in the region and also on the strategic stability of the South Asia which can lead to minor military retaliation. But even during the heightened tensions internally and externally, Pakistan has always responded as a mature and responsible nuclear state.[80]

Najm Us Saqib, a former Pakistani ambassador analyzed in a 6 June article that while the most prudent option for Pakistan was co-existence with India and a focus on internal development, it must also develop capabilities for proportionate responses to Indian provocations and not just depend on a response based on restraint.

> The latest action that showed New Delhi's intentions on having or not having any discussion on Kashmir or the nature of relations with Pakistan was India's "accidental" firing of a version of the BrahMos cruise missile into Pakistan on March 9. Guessing and calculating another provocative "adventure" from India seems to be a futile exercise unless both sides reach a kind of "understanding" on not crossing the "red-line"....In any case, after experiencing the "accidental" firing of a missile from India, it was time for Pakistan to draw a line and keep an adequate and proportionate response handy, just in case. There must be a limit to the doctrine of restraint.[81]

Nadir Ali of the Institute for Strategic Studies, Islamabad, focused on the institutional structure of India's nuclear policymaking in an article on 16 June and pointed to the alleged cover-up of the missile misfire incident by the Indian military.

> there is minimal open source material accessible on the decision-making process of India's nuclear command and control. This missile was not a nuclear missile but a strategic missile – any non-nuclear strategic missile should likewise be controlled by the Nuclear Command Authority (NCA). One must need to understand the internal bureaucratic dynamics within India, where the military is completely excluded from decision-making in the NCA. The delivery systems are largely under the control of the Defence Research and Development Organization (DRDO), while the warheads are under the control of the Department of Atomic Energy (DAE). Therefore, concerning the strategic problem, India has always

> claimed that all missiles will be regulated or authorised by political authority.In this case, one would presume that the political leadership was notified a day later because it took two days for India to openly recognise that a missile was fired. Despite the fact that whatever happened on the base has been covered up by the Indian military, which is why no explanation has been provided.[82]

Three Causative Scenarios: Advertent Move, Accidental Launch, Unauthorized Launch

Rizwana Abbasi, Associate Professor at Department of International Relations in National University of Modern Languages, Islamabad, and Muhammad Saeed Uzzaman, Lecturer at Department of International Relations in National University of Modern Languages, Rawalpindi, authored a paper which relied on primary and secondary data sources and advanced a conceptual framework constructed around three causative scenarios to warn about the high risk of conflict from a future accidental launch given the lack of Indian communication through the military (DGMO) and political (Foreign Secretary) hotlines.

> Three possible scenarios help draw plausible findings on this incident and each one brings a set of troubling questions to the spotlight. One likely scenario is that the Indian missile landing during peacetime was an advertent move to test the readiness of Pakistan's defense systems, probing the latter's monitoring and tracking systems and its response level, precipitating a crisis....The second scenario is that it was indeed a malfunction involving human and technical errors. One, India revealed that the accidental launch of the missile took place during a military exercise supervised by a Directorate of Air Staff Inspection (DASI) at one of the Indian Air Force (IAF) bases located in the north of India. This could be possible since India's supersonic missile development is still incomplete, and its accuracy is indeed not as robust as expected. Lapses in the Indian missile systems and their lack of accuracy are largely apparent. The third scenario is that it was an unauthorized missile launch. The reports suggest that the IAF personnel are involved in the missile incident. Even one of the (Indian National) Congress leaders raised questions on the involvement of the IAF personnel in the missile launch and asked the relevant authorities of the IAF to share the requisite details of how and why its personnel fired the missile. According to the Indian media reports, the IAF investigation of the technical standard operating procedures points towards the Group Captain for firing an unarmed BrahMos Supersonic Missile into Pakistan's territory... Regardless of whether it was an advertent, accidental, or unauthorized launch, by all dimensions, the incident was exceedingly risky and

> carried the enormous potential to invite a more significant crisis, thus endangering humanity.[83]

Ghazala Jalil, Research Fellow at the Arms Control and Disarmament Centre, at the Institute for Strategic Studies, Islamabad (ISSI) analyzed possible Indian motives (inadvertent versus deliberate) behind the missile misfire in a 6 July issue brief. She also joined other Pakistani analysts in recommending that cruise missiles also be brought under the ambit of the Ballistic Missile Flight Test Pre-Notification Agreement of 2005.

> Another disturbing aspect of the incident was a 48-hour delay in any Indian acknowledgement of the incident or communication with Pakistan. There is an army-to-army hotline that is not used....If this was indeed an accidental launch, then India needs to look into its command and control systems and adopt fail-safe launch protocols because a nuclear South Asia cannot afford missile launches – inadvertent or deliberate. If this was not an accidental launch then what could be Indian motives? It may have been an attempt to test Pakistan's deterrence capability and its resolve to respond to missile launches. The proposal for a hotline for nuclear-related issues has been on the table for almost two decades. This must be negotiated before the region faces the next crisis. The two countries already have a ballistic-missile flight test pre-notification agreement, cruise missiles can be added to this.[84]

Dr. Naeem Salik, Executive Director at the Strategic Vision Institute, Islamabad reckoned in a paper in *Springer Proceedings in Physics* that "Armageddon" was only prevented by a "combination of good luck, and cool-headed Pakistani response and a relatively calm environment prevailing between the two countries." He also worried about other eventualities that could have led to an escalatory cycle in the aftermath of the missile accident.

> *What if the missile had landed in a thickly populated area or hit a sensitive target?* In both cases public sentiment would have enraged demanding a reprisal by Pakistani government and in case of a sensitive target in addition to the public outrage need would have been felt by the Pakistani military leadership to payback for the damages caused by the missile.
>
> *What if it had hit a passenger airliner and caused large scale civilian casualties?* Depending on whether it was international flight operated by some foreign country or a domestic flight, and whether it involved a small airliner or a large aircraft with large number of passengers on board the severity of the incident would have varied but in all eventualities the connotations would have been very severe and long lasting.

> *What if Pakistan had immediately responded in kind?* If Pakistan had acted on the basis of worst-case scenario suspecting the missile to be carrying a lethal payload and in accordance with its declared policy of 'quid-pro-quo' plus India would have found itself in a bind; whether to accept its earlier mistake and Pakistan's legitimate right to act in self defence and refrain from further action or react to Pakistani response and pave the way for rapid escalation.
>
> *What if the missile was armed with a conventional, worse still a nuclear warhead?* Had the incoming Indian missile been armed with a conventional warhead it would have caused much greater damage on ground compared with what it actually did and if it was armed with a nuclear warhead, it would have undoubtedly caused a nuclear war between two nuclear armed states.[85]

Pakistan's Restraint, India's Hindutva Extremism, Hegemonic Aspirations, Risk Reduction

Salma Malik, Assistant Professor at the Department of Defence & Strategic Studies, Quaid-i-Azam University, also praised Pakistan's prudence and restraint for preventing an escalation, criticized the international community for its apathy towards the missile accident and castigated India for scapegoating a mid-ranked officer in its Court of Inquiry.

> With the world focused on the Ukrainian conflict, not much attention was given to this grave occurrence. Alternatively, had this act been done by Pakistan and not India, would the world still be so nonchalant about the entire episode?....Ever since the Modi government has been voted into power, it has made no qualms about its belligerent and aggressive posturing with regards to Pakistan....Indian policy maker's assurance of an inquiry and placing the blame conveniently on a mid-rank officer in charge, is again an arrogant dismissal of the big picture. This action brought the region and correspondingly the world closer to a nuclear midnight. Especially when New Delhi not only plans revising its No First Use option, which in any case is very ambiguously framed, but also has a portion of its nuclear force, particularly those systems that are—designed for use against Pakistan, now kept at a high state of readiness, capable of being operationalised and released within seconds or minutes in a crisis. Should Pakistan then consider this as a test run by India to gauge the threshold of its deterrence response scenario as well as explore gaps in its operational readiness.[86]

In a 28 August article for *Pakistan Observer*, retired Group Captain Sultan M Hali reminded of India's multiple failed previous missile tests, criticizing them as the "irresponsible behavior" of a country with hegemonic aspirations in South Asia.

> Missile disasters are not new in India. In November 2015, a major aviation disaster was in the offing, when Indian Navy test fired several lethal surface-to-surface missiles above the Arabian Sea as the airspace remained open to air traffic, horrifying the world. In a fit of frenzy over its debacle against Chinese troops in October 2020, India test fired ten different missile systems. The bid to cow down China could have easily backfired and caused a serious accident. The list of failed Indian missiles is long but no lessons have been learnt by the jingoistic and irresponsible South Asian wannabe power. Agni II failed to reach target thrice during the testing phase; Agni III and IV too failed tests. BrahMos failed initial tests. 'Nirbhay' was unsuccessful in acquiring its target. India's irresponsible behaviour cannot be condoned.[87]

Haris Akmal, an Islamabad-based researcher, echoed other Pakistani analysts in his 1 September column for *PakistanToday* in which he suspected that the missile launch was a deliberate Indian move to test for vulnerabilities in Pakistan's air defense system. Akmal based his contention on two observations; India's non-utilization of the DGMO hotline in the immediate aftermath of the incident and the self-destruct feature on the missile not being activated once it went astray.

> First, if it was accidental, why did India keep silent after the incident occurred? India could have activated the 'hotline' between the respective Directors Generals of Military Operations in Delhi and Islamabad. The basic purpose of the hotline has failed. It was meant to calm a provocative security situation with advance communication. Secondly, India's response has been vague and unconvincing. One wonders why the self-destruct feature of the missile wasn't activated before entering into Pakistan. Can Pakistan be blamed if it assumes that some rogue elements have taken control of India's command and control system?...In fact, there is a genuine sense in Pakistan that India seems not interested in a serious overview of its command and control mechanism.[88]

A particularly important contributor to the Pakistani strategic discourse was Lt Gen (retired) Khalid Kidwai, adviser to Pakistan's National Command Authority (NCA), former Director General of its Strategic Plans Division and a 47-year career veteran of the Pakistan Army. Speaking at the CISS-IISS 8th International Workshop on Strategic Stability in South Asia organized by the Islamabad-based, Center for International Strategic Studies (CISS) on 21 September, Kidwai worried about the rise of a "toxic mix of hardcore and ruthlessly aggressive Hindutva fundamentalist ideology" which had "permeated Indian society, governance and the diaspora in the West."

He warned that the "extremist, fundamentalist leadership" possessing the "custodial control of India's large triad of nuclear arsenal" posed a "serious risk to strategic stability in South Asia," recounting the Balakot air strike of February 2019 as the "intoxication" of extremists at the "highest levels of the BJP leadership." Commenting more pointedly on the missile misfire, Kidwai confidently analyzed that it was a deliberate attempt to test Pakistan's air defense alert levels and operational responses.

> the now BJP-oriented military launched deep inside Pakistan a nuclear-capable BrahMos missile with a controlled trajectory, pre-planned vertical and horizontal waypoint coordinates, fed into the onboard guidance and control computer along with the geographic coordinates of the launch point and the target. The objective was very clear, to test Pakistan's air defence alert levels and operational responses. India did not particularly care that a missile posed a destructive threat for some 7 to 8 minutes to at least a dozen commercial airlines in the air at that time. I would like to state with complete responsibility and confidence that the launch was no accident as claimed slyly by India. The launch could not have taken place without political clearence at the highest level and detailed planning over a number of days and weeks in the military chain of command to include technical preparations of the missile, the missile launcher, storage and ground-deployment drills with full involvement of the immediate missile launch crew of at least 10 to 15 odd personnel besides the hundreds of other personnel that comprise a BrahMos strategic missile group.[89]

Drawing from his 15 year experience of having overseen 50 missile tests, Kidwai rubbished the Indian claim of an accidental launch and praised Pakistan for its restrained response. He cast aspersions over the results of the Indian Court of Inquiry, and characterized the three dismissed IAF officers as the fall guys of an Indian cover-up operation. He also warned against mistaking Pakistani restraint for weakness.

> I have lived, breathed and conducted an array of strategic, ballistic and cruise missiles for 15 years as DG-SPD perhaps over 50 in number. I know my missiles and missile technology. I know the sequence of technical preparations, the sequence of deployment drills and the SOP's that just have to be followed without which a missile launch can never take place. Nuclear capable ballistic and cruise missiles are highly complex state of the art machines with meticulously detailed and controlled launch procedures and SOP's including the capability to self-destruct if things go wrong. These do not fire off accidentally like infantry rifles during

> weapons cleaning drills. India put out a well-rehearsed though laughable cover story of an accidental launch which is technically and operationally speaking entirely mischievous to pull wool over our eyes and the worlds. No serious professional will buy this silly story. Three IAF officers have reportedly been made the fall guys for what essentially was a military operation conducted against Pakistan. I recommend strongly that India look after them and compensate them well for their silence. On both occasions, at Balakot 2019 and Shorekot 2022, Pakistan displayed restraint and maturity in diffusing the irresponsibly generated tensions thereby preventing South Asia from spiralling into potential catastrophes. It will be foolhardy to take Pakistan's restraint and maturity as a weakness and continue to test the limits of strategic stability. Perhaps India is attempting to establish a pattern of incidents over time as a strategy in order to desensitize the international community for their future operations.[90]

Amber Abid, Research Associate at the Strategic Vision Institute, Islamabad, reiterated the familiar reasons for Pakistani suspicions that the Indian launch was a deliberate ploy in a 10 April article published by the Rawalpindi-based *Global Defense Insight,* a Pakistani think-tank.

> The supersonic missile traveling at such a higher speed can't take 90 degrees sudden turn provided its fast speed of Mach 3. Furthermore, the timing it was launched sparks concerns around their termed position of technical maintenance; as such matters are done during the day, not in the twilight. Despite having the 2004 hotline agreement between the two countries, the so-called accidental launch was not been intimated.... As far as Indian intentions are concerned, whether it was accidental or intentional, considering the above facts and Indian aggressive counterforce posture and aggressive designs against Pakistan, this could be an attempt to assess Pakistan's deterrence capabilities and unclear response measures. India wants to undermine Pakistan's deterrent capabilities.[91]

Moiz Khan is a Researcher at the Centre for Aerospace & Security Studies (CASS), Islamabad, made the case for Confidence Building Measures (CBM's) and Risk Reduction Measures (RRM's) in an 6 October article.

> India and Pakistan could also consider devising new Risk Reduction Measures (RRMs). For example, missiles that are scheduled to be inspected, both countries need to configure their weapons' guidance systems to unoccupied places such as oceans or deserts where they pose minimum dangers. Moreover, the weapons' pre-fed adversary target locations need to be removed while used for inspection, training, or simulated exercises. The maintenance of actual coordinates of adversary

targets could lead to unintended escalation in accidental launches. These measures would not only help avoid accidents, they could also serve as an added layer of protocol to minimise the possibility of unauthorised launch. However, accidents happen despite best safety protocols as there are limits of safety procedures. In such a possibility, there is a need of haste to communicate accidental launches.[92]

Dr. Shahid Hameed, Director of Research at the Center of International Strategic Studies (CISS) AJK, called for the international community and International Atomic Energy Agency (IAEA) to examine India's nuclear command and control system.

> to prevent a global crisis, the international community and the International Atomic Energy Agency (IAEA) must reexamine the Indian nuclear strategy and the recent controversial discussions of revisions to NFU (No First Use Policy). The IAEA needs to examine and suggest updates to India's nuclear command and control system and should consider India's nuclear posture as a serious issue. On the other hand, Pakistan must offer a framework by involving the other stakeholders in the region, including China and Russia, to regulate and oversee India's aggressive nuclear approach.[93]

Commenting in a 15 December article for the *Pakistan Observer,* independent researcher Syed Qamar Afzal Rizwi took issue with the Indian media's alleged misreporting of the IAEA's position on the missile accident.

> Currently Pakistan's Foreign Office has deemed Indian media's report a "disingenuous attempt" for misreporting IAEA's Rafael Mariano Grossi's response to a query about the firing of India's missile into Pakistan (March 9, this year). Irresponsibly and mischievously, the Indian media reported that the BrahMos nuclear capable missile fired into Pakistan's territory from India (March 9) was not a cause for any specific concern for the UN's nuclear watchdog. Pakistan is highly justified in arguing that the IAEA's Director General response by no means can be "purposely misinterpreted" to "trivialise" the incident of a nuclear-capable BrahMos missile fire with grave implications for regional and global security in that the issue has raised logical questions about India's conduct as a nuclear state including, whether it was actually an accident.[94]

Indian Strategists—Accident-Centric Focus, Limited Contextual Analysis

The discussion on the missile accident in the Indian strategic community occurred in the same environment of information deficit as the discussion in the Indian media (covered in Chapter 2). Indian strategists grappled with the semantic details of terse government statements, their chronology and selective anonymous leaks to the media by sources close to the government to construct details of the accident such as the type of missile involved, possible causes of the accident and the most recent developments pertaining to India's Court of Inquiry. Even as strategists engaged in frank discussions on the causes and consequences of the accident, they also walked the tightrope of tempering their objective analyses of the ramifications of the missile accident with nuanced policy recommendations and contextualizing observations which ultimately supported the Indian government's Court of Inquiry and India's image as a responsible state.

The discussions in the Indian strategic community were mostly accident/event-centric, focusing on the details of the accident and viewing it for most part as a mishap which required the reactivation of existing crisis management mechanisms (hotlines) and reform of standard operating procedures. The discussions in the Indian strategic community overlapped with but also diverged from the discussions in the Indian media, given the greater range of issues discussed and the greater technical depth of the discussion.

In the days following the accident, Indian strategists attempted educated guesses regarding the exact type of missile (modified ISRO rocket, BrahMos Air Launched Cruise Missile) that had intruded Pakistan's airspace. They also closely studied the Indian government's two official statements on 11 and 15 March respectively in order to ascertain whether the missile was accidentally launched in the course of "routine maintenance" or whether it was discharged inadvertently during "routine maintenance and inspection."

Some strategic experts also attempted to identify the specific locus of the error leading to the accident and ventured that it had occurred in the firing sequence of the missile due to a major violation of standard operating procedures. Specific causes such as the entry of incorrect target coordinates, cyber jamming of the missile in mid-flight by Pakistan and corruption of the target coordinates fed into the missile were also discussed.

Also focused on were important issues such as whether the accident occurred due to the lack of adequate protocols or due to lack of adequate adherence to existing procedures. Strategists also dwelled on whether the BrahMos had a self-destruct mechanism and the reasons for the inability of the missile personnel to activate it including a possible delay in orders from higher authorities, a failure of the mechanism after it was activated, lack of knowledge on the part of the missile unit on how to activate the mechanism or the mechanism not being capable of being activated as the missile was out of range.

Fears were also expressed of Pakistani infiltration into India's missile complex and its suborning of Indian missile personnel to sabotage Indian missile systems. There were different views regarding whether India had used the military hotline to inform Pakistan in the hours after the accident. There was also genuine puzzlement as to why Pakistan had not shot down the missile with views ranging from Pakistan not having detected the missile at all to it having displayed restraint in detecting the missile but choosing not to respond. Many Indian strategists were frank in their praise for what they saw as Pakistan's mature reaction to an intruding supersonic missile in a situation where it had no idea whether the missile was armed with a warhead or not.

At least one member of the strategic community also relayed a fleeting glimpse of the skepticism of the Indian armed forces regarding the technical malfunction rationale initially advanced by the Indian government. This skepticism of the armed forces, when combined with the reported dissatisfaction among the designers of the BrahMos following the government's identification of the technical malfunction rationale (covered in Chapter 1), highlights the role of another key actor in the discourse on the Indian side.

Strategists also excoriated the Indian government for its 48-hour delay in acknowledging the accidental launch and raised questions regarding the skilling and competence of supposedly "second-rung" missile operators in an air force where top performers were preferred for fighter pilots and "broken wings" assigned to missile units. Even as they criticized the Indian government on multiple levels ranging from the delay in acknowledging the accident to the absence of the right standard operating procedures, some strategists also made an attempt to shore up confidence in India's command and control system by pointing out that the BrahMos cruise missile had a one-man protocol versus the two-man protocol for India's ballistic missiles. Even as they forcefully enjoined the government to conduct a timebound, impartial probe to affix responsibility, they also dismissed any chance of a joint probe or the publicizing of the full report of the Indian Court of Inquiry on grounds of national security.

A broader attempt at contextualization was also made by pointing to multiple missile accidents internationally, including between the US and the Soviet Union. Finally, there was across-the-board recognition of the risk posed by the accident to South Asian stability but the policy recommendations were limited to the strengthening of bilateral crisis management mechanisms such as the military hotline and to the continued separation of conventional and nuclear weapons systems to avoid entanglement. Missing was a discussion on measures to ensure prompt communication within the Indian system between missile personnel, the military top-brass and the political authorities in case of a similar, future missile accident. Given the dominant framing of the missile accident as a mishap involving a cruise missile which could only carry a conventional (and not also a nuclear) warhead, the discussion focussed only to a limited extent on the ramifications of the accident for nuclear deterrence, strategic signaling, escalation and de-escalation in South Asia.

Only a minority in the Indian strategic community claimed the missile launch as a deliberate Indian act which was part of a more assertive Indian posture to assess Pakistani operational readiness.

Pakistani Strategists—Context-Centric Analysis with Accident as a Key Indicator

The discourse by Pakistani strategists on the accident was more oriented towards viewing it in the context of India's current right-wing government, the extremist actions that were perceived to result from its ascribed Hindutva ideology and the systemic problems with the command and control mechanisms of India's strategic weapons systems. An important section of the Pakistani foreign policy elite (including Lt. Gen. (Retd) Khalid Kidwai and former Director General of the Strategic Plans Division Force) viewed the BrahMos missile as nuclear capable (dual-use) and its launch being a deliberate attempt by a Hindu extremist government to test Pakistani air defenses and its operational readiness. The missile intrusion was also viewed by other strategists as an attempt to normalize cross-border strikes as part of a shift away from (and weakening of) India's No First Use (NFU) doctrine to a more flexible counterforce posture. The inference drawn from the missile accident was that Indian missiles were in launch-ready mode. Strategists differed on Pakistan's response to the intruding missile with some commending Pakistan for acting with restraint and maturity while others criticizing the Pakistani Air Force for not attempting to shoot down a missile from an adversary nation which could have potentially carried a nuclear warhead.

Given the views of some strategists that Pakistan would have retaliated with its own missile had it hit a population center or a military facility as part of its 'quid pro quo plus' strategy and the aforementioned severe trust deficit concerning India's motives, any future missile intrusion (whether accidental or otherwise) is likely to significantly lower the threshold for a Pakistani response. Pakistani strategists also called for discussions on a range of proportionate Pakistani responses to any perceived provocations from the Indian side and any finalization of such responses is also likely to lower the threshold for Pakistani retaliation.

The dual-use perception of the BrahMos missile also led Pakistani strategists to worry about the reliability of India's nuclear command and control systems and the competence and morale of its missile personnel. They also cast doubt on the communication linkages within the Indian establishment between its launch-authorizing political authority, military top brass and on-ground missile personnel (especially in an environment where factions within the Indian military were pushing in support of continuance of the existing system of Corps versus the new system of integrated theaterization). Strategists also weighed in on the inability of the Indian DRDO to successfully absorb Russian technological assistance made available under the BrahMos joint venture.

In addition to the possibility of a software glitch, incorrect entry of coordinates and the lack of training and skilling of Indian missile unit personnel as causes for the accident, Pakistani strategists were also alarmed at the prospect of the erosion of civilian primacy in the decision-making chain and the unauthorized launch of the missile by a either a disgruntled/extremist-minded military insider.

All the aforementioned Pakistani threat perceptions played a role in the Pakistani government's demand for a joint probe into the accident. They also led to intense Pakistani skepticism and rejection of the findings of India's Court of Inquiry and its publicly announced decision to dismiss three IAF officers. The diagnosis of the causes of the missile accident as a 'whole-of-system' problem by Pakistani strategists led to their dismissal of the Indian probe's result as a cover-up by the Indian military to hide systemic issues in India's command and control systems by making scapegoats of the three officers.

Pakistani strategists also came up with a raft of solutions to minimize the possibility of a missile accident in the future including the inclusion of cruise missiles in the 2005 Agreement on Pre-Notification of Ballistic Missile Tests, political dialogue on all issues including Kashmir, reactivation of the DGMO military hotline, discussions to clarify the alert status of missiles and the ones assigned conventional versus nuclear-strategic roles.

Notes

1 Rajeshwari Pillai Rajagopalan, "The Many Unanswered Questions About India's Missile Accident," *Observer Research Foundation*, March 25, 2022, accessed February 2, 2023, https://www.orfonline.org/research/the-many-unanswered-questions-about-indias-missile-accident/

2 Ruhee Neog, "An Accidental Missile Launch and a Lesson for Indian Communications," *Asia-Pacific Leadership Network*, April 29, 2022, accessed February 2, 2023, https://www.apln.network/analysis/commentaries/an-accidental-missile-launch-and-a-lesson-for-indian-communications
Imran Siddhiqui, "Missile Error, Communication Lapse: War Veterans," *The Telegraph*, March 17, 2022, accessed February 3, 2023, https://www.telegraphindia.com/india/missile-error-communication-lapse-say-war-veterans/cid/1856313

3 Rashme Sehgal, "'Even If Authorised, Missile Firing Makes No Sense.'" *Rediff*, March 13, 2022, accessed February 3, 2023, https://www.rediff.com/news/interview/dr-bharat-karnad-even-if-authorised-missile-firing-makes-no-sense/20220313.htm

4 Imran Siddhiqui, "Missile Error, Communication Lapse: War Veterans," *The Telegraph*, March 17, 2022, accessed February 3, 2023, https://www.telegraphindia.com/india/missile-error-communication-lapse-say-war-veterans/cid/1856313

5 Ajai Shukla, "BrahMos Cruise Missile Cannot Be Mistaken for a Nuclear-Tipped Missile," *Business Standard*. March 15, 2022, accessed February 4, 2023, https://www.ajaishukla.com/2022/03/brahmos-cruise-missile-cannot-be.html
6 Snehesh Alex Philip, "Accidentally Fired Missile into Pakistan Due to Tech Glitch, Says India. 'It Was BrahMos,'" *ThePrint*, March 11, 2022, accessed February 4, 2023, https://theprint.in/defence/accidentally-fired-missile-into-pakistan-due-to-tech-glitch-says-india-it-was-brahmos/869387/
7 Prasanna Zore, "'Why Would India Deliberately Fire an Unarmed Missile into Pakistan?'" *Rediff*, March 15, 2022, accessed February 4, 2023, https://www.rediff.com/news/interview/dr-yogesh-joshi-why-would-india-deliberately-fire-an-unarmed-missile-into-pakistan/20220315.htm
8 Happymon Jacob, "A Misfiring and Its Trail of Poor Strategic Stability," *The Hindu*, March 15, 2022, accessed February 4, 2023, https://www.thehindu.com/opinion/lead/a-misfiring-and-its-trail-of-poor-strategic-stability/article65228306.ece
9 PTI, "Every Child in India Will Be Safe, Educated by 2047: Kailash Satyarthi," *Deccan Herald*, May 3, 2022, accessed February 5, 2023, https://www.deccanherald.com/india/every-child-in-india-will-be-safe-educated-by-2047-kailash-satyarthi-1105992.html
10 Rahat Iqbal, "India's Brahmos Faux Pas and International Response," *Centre for Strategic and Contemporary Research*, May 18, 2022, accessed February 5, 2023, https://cscr.pk/explore/themes/defense-security/indias-brahmos-faux-pas-and-international-response/
11 Faizan Fakhar, "BrahMos Misadventure: Accidental or Deliberate?," *The Express Tribune*, March 23, 2022, accessed February 6, 2023, https://tribune.com.pk/story/2349154/brahmos-misadventure-accidental-or-deliberate
12 Staff Reporter, "Indian Missile Incident Questions Its Arsenal Credibility: Experts," *DAWN.COM*, April 23, 2022, accessed February 6, 2023, https://www.dawn.com/news/1686298
13 Ghazala Yasmin Jalil, "India's Accidental Missile Launch--Implications and Challenges Ahead," *Institute of Strategic Studies-Islamabad*, July 6, 2022, accessed May 12, 2023, IB_Ghazala_July_6_2022.pdf (issi.org.pk), pp. 3–5.
14 Flashpoint, "India's 'Accidental' Missile Launch," *Jinnah Institute*, March 25, 2022, accessed February 9, 2023, https://jinnah-institute.org/feature/indias-accidental-missile-launch/
15 Ahmed Saeed, "India's Missile 'Mistake' and Global Bias," *T Magazine*, March 27, 2022, accessed February 9, 2023, https://tribune.com.pk/story/2349804/indias-missile-mistake-and-global-bias
16 CISS Pakistan, "CISS Webinar on 'Indian Missile Fiasco: Technical Malfunction and Failure of Diplomacy,'" *Center for International Strategic Studies*, April 21, 2022, accessed February 9, 2023, https://ciss.org.pk/ciss-webinar-on-indian-missile-fiasco-technical-malfunction-and-failure-of-diplomacy/
17 Rahul Bedi, "India Needs to Come Clean on 'Accidental Firing of Missile' to Avert Conflict with Pakistan," *The Wire*, March 12, 2022, accessed February 10, 2023, https://thewire.in/security/india-needs-to-come-clean-on-accidental-firing-of-missile-to-avert-conflict-with-pakistan.
18 Brig Kuldip Singh (Retd), "Indian Missile in Pakistan: What Led to the 'Accidental Firing'?" *TheQuint*, March 12, 2022, accessed February 10, 2023, https://www.thequint.com/opinion/indian-missile-in-pakistan-what-led-to-the-accidental-firing#read-more
19 Bharat Karnad, "Criminal Laxity or Is It Complicity in the 'Misfiring' of the Brahmos Missile? Strategic Dimensions (Augmented)," *Security Wise (blog)*,

March 12, 2022, accessed February 11, 2023, https://bharatkarnad.com/2022/03/12/criminal-laxity-or-is-it-complicity-in-the-misfiring-of-the-brahmos-missile/
20 Rashme Sehgal, "'Even If Authorised, Missile Firing Makes No Sense.'" *Rediff*, March 13, 2022, accessed February 12, 2023, https://www.rediff.com/news/interview/dr-bharat-karnad-even-if-authorised-missile-firing-makes-no-sense/20220313.htm
21 Sood Rakesh, "About," October 11, 2020, accessed February 11, 2023, https://rakeshsood.in/about/
22 Seema Guha, "India's Accidental Firing of Unarmed Missile into Pakistan Raises Questions of Safety Protocols," *Outlook*, March 13, 2022, accessed March 13, 2023, https://www.outlookindia.com/national/india-accidental-firing-of-missile-into-pakistan-questions-of-safety-protocols-news-186688?utm_source=related_story
23 "Manoj Joshi - Orf," *ORF*, accessed March 13, 2023, http://20.244.136.131/people-expert/manoj-joshi
24 TOI, "The Case of the Runaway Missile," *The Times of India*, March 14, 2022, accessed February 13, 2023, https://timesofindia.indiatimes.com/podcasts/the-times-of-india-podcast/the-case-of-the-runaway-missile/toipodcast/90188990.cms
25 TOI, "The Case of the Runaway Missile," *The Times of India*, March 14, 2022, accessed February 13, 2023, https://timesofindia.indiatimes.com/podcasts/the-times-of-india-podcast/the-case-of-the-runaway-missile/toipodcast/90188990.cms
26 Tara Kartha, "Why Near-Disaster BrahMos Accident Is an Opportunity for India & Pakistan to Discuss Safety," *ThePrint,* March 14, 2022, accessed February 13, 2023, https://theprint.in/opinion/why-brahmos-accident-is-opportunity-for-india-pakistan-to-discuss-safety/871739/
27 Tara Kartha, "Why Near-Disaster BrahMos Accident Is an Opportunity for India & Pakistan to Discuss Safety," *ThePrint,* March 14, 2022, accessed February 13, 2023, https://theprint.in/opinion/why-brahmos-accident-is-opportunity-for-india-pakistan-to-discuss-safety/871739/
28 Manan Bhatt, "The Curious Case of a Misfired Missile," *DNA India*, March 14, 2022, accessed February 13, 2023, https://www.dnaindia.com/analysis/report-the-curious-case-of-a-misfired-missile-india-pakistan-2939848
29 Nidhi Razdhan, "How Did India Accidentally Fire a Missile Into Pakistan?" *NDTV*, March 14, 2022, accessed February 15, 2023, https://www.youtube.com/watch?v=DqCT8QONTVw
30 Ajai Shukla, "BrahMos Cruise Missile Cannot Be Mistaken for a Nuclear-Tipped Missile," *Business Standard,* March 15, 2022, accessed February 15, 2023, https://www.ajaishukla.com/2022/03/brahmos-cruise-missile-cannot-be.html
31 ORF, "About Rajeshwari Pillai Rajagopalan," *Observer Research Foundation*, accessed February 15, 2023, http://20.244.136.131/people-expert/rajeswari-pillai-rajagopalan
32 Karan Thapar, "Missile Accident Raises Disturbing Questions, Dents India's Image As Responsible Nuclear State," *The Wire*, March 15, 2022, accessed February 15, 2023, https://www.youtube.com/watch?v=W0DnDWrnqSg
33 Happymon Jacob, "A Misfiring and Its Trail of Poor Strategic Stability," *The Hindu,* March 15, 2022, accessed February 15, 2023, https://www.thehindu.com/opinion/lead/a-misfiring-and-its-trail-of-poor-strategic-stability/article65228306.ece
BR Webdesk-Reuters, "India Says It 'accidentally' Fired Missile into Pakistan Due to 'Technical Malfunction,'" *Business Recorder,* March 11, 2022, accessed February 16, 2023, https://www.brecorder.com/news/40160162

34 Manvendra Singh, "IAF Has a Lot to Answer on BrahMos Firing. It Has Dented India's Image of Being Responsible," *ThePrint*, March 16, 2022, accessed February 18, 2023, https://theprint.in/opinion/iaf-has-a-lot-to-answer-on-brahmos-firing-it-has-dented-indias-image-of-being-responsible/874765/
35 Barkha Dutt, "India's Missile Misfire & Pakistan's near Retaliation l Inside the Close Shave," *Mojo Story*, March 16, 2022, accessed February 23, 2023, https://www.youtube.com/watch?v=_pTvqv275ts
36 D. Raghunandan, "The Curious Incident of the 'Misfired' Missile," *NewsClick,* March 17, 2022, accessed February 23, 2023, https://www.newsclick.in/curious-incident-misfired-missile
37 Imran Siddhiqui, "Missile Error, Communication Lapse: War Veterans," *The Telegraph*, March 17, 2022, accessed February 24, 2023, https://www.telegraphindia.com/india/missile-error-communication-lapse-say-war-veterans/cid/1856313
38 Imran Siddhiqui, "Missile Error, Communication Lapse: War Veterans," *The Telegraph*, March 17, 2022, accessed February 24, 2023, https://www.telegraphindia.com/india/missile-error-communication-lapse-say-war-veterans/cid/1856313
39 Imran Siddhiqui, "Missile Error, Communication Lapse: War Veterans," *The Telegraph*, March 17, 2022, accessed February 24, 2023, https://www.telegraphindia.com/india/missile-error-communication-lapse-say-war-veterans/cid/1856313
40 ANI, "Pakistan's 'missile' Fails to Reach Target, Crashes in Sindh," *ThePrint,* March 18, 2022, accessed February 25, 2023, https://theprint.in/world/pakistans-missile-fails-to-reach-target-crashes-in-sindh/878464/
41 Krishn Kaushik, "Explained: When a Missile Misfires," *The Indian Express,* March 19, 2022, accessed February, 25 2023, https://indianexpress.com/article/explained/explained-when-a-missile-misfires-india-pakistan-brahmos-7815962/
42 ORF, "About Rajeshwari Pillai Rajagopalan," *Observer Research Foundation*, accessed February 27, 2023, http://20.244.136.131/people-expert/rajeswari-pillai-rajagopalan
43 Rajeshwari Pillai Rajagopalan, "The Many Unanswered Questions About India's Missile Accident," *Observer Research Foundation*, March 25, 2022, accessed February 27, 2023, https://www.orfonline.org/research/the-many-unanswered-questions-about-indias-missile-accident/
44 Manish Tewari, "Manish Tewari | Unguided Missile: A Case for Indo-Pak N-Dialogue," *The Asian Age,* March 26, 2022, accessed February 27, 2023, https://www.asianage.com/opinion/columnists/260322/manish-tewari-unguided-missile-a-case-for-indo-pak-n-dialogue.html
45 Lt Gen (Retd) D. S. Hooda, "A Distant War and Some Home Truths," *Deccan Herald,* July 15, 2023, accessed November 29, 2023, https://www.deccanherald.com/specials/a-distant-war-and-some-home-truths-1237149.html
46 CSDR-New Delhi, "Skirting the Nuclear Threshold: Conflict De-Escalation and Strategic Stability in South Asia. Council for Strategic and Defense Research," March 31, 2022, accessed February 27, 2023, https://www.youtube.com/watch?v=65h66CFQv1w
47 Ruhee Neog, "An Accidental Missile Launch and a Lesson for Indian Communications," *Asia-Pacific Leadership Network*, April 29, 2022, accessed March 2, 2023, https://www.apln.network/analysis/commentaries/an-accidental-missile-launch-and-a-lesson-for-indian-communications
48 KP Sanjeev Kumar, "BrahMos 'Misfire': Indian Govt's Dismissal of 3 Officers Raises Key Questions," *TheQuint,* August 25, 2022, accessed March 2, 2023,

https://www.thequint.com/opinion/brahmos-misfire-indian-govts-dismissal-of-3-officers-raises-key-questions

49 KP Sanjeev Kumar, "BrahMos 'Misfire': Indian Govt's Dismissal of 3 Officers Raises Key Questions," *TheQuint,* August 25, 2022, accessed March 2, 2023, https://www.thequint.com/opinion/brahmos-misfire-indian-govts-dismissal-of-3-officers-raises-key-questions

50 T. K. Chatterjee, "The Brahmos Misfire," *Financial Express*, August 29, 2022, accessed March 2, 2023, https://www.financialexpress.com/business/defence-the-brahmos-misfire-2648120/

51 Qamar Cheema, "Major Gaurav Arya Talks on Brahmos Missile Inquiry: Why PM Modi Does Not Respond Pak Leadership," *Dr. Qamar Cheema Show,* August 27, 2022, accessed March 2, 2023, https://www.youtube.com/watch?v=SLZfWlkap1w

52 TNN, "Pakistan's Pathankot Probe Team, with ISI Official, Reaches India," *Times of India,* March 27, 2016, accessed March 5, 2023, https://timesofindia.indiatimes.com/india/pakistans-pathankot-probe-team-with-isi-official-reaches-india/articleshow/51570076.cms

53 Qamar Cheema, "Major Gaurav Arya Talks on Brahmos Missile Inquiry: Why PM Modi Does Not Respond Pak Leadership," *Dr. Qamar Cheema Show,* August 27, 2022, accessed March 5, 2023, https://www.youtube.com/watch?v=SLZfWlkap1w

54 "DGISPR Demands Explanation after Indian Projectile Crashes near Mian Channu," *ARY News,* March 11 2022, accessed March 6, 2023. https://www.youtube.com/watch?v=qhy9GsZl0Tk

55 Qamar Cheema, "Pakistan Needs S 400 System after Missile Incident:Political Consensus Needed 4 Defense Procurement," *The Qamar Cheema Show,* March 12, 2022, accessed March 10, 2023, https://www.youtube.com/watch?v=GGj9FRM5ja0

56 "BrahMos Supersonic Missile - Bharat Ne Yeh Missile Ghalti Se Chalaya Ya Jaan Boojh Kar?" - *SAMAA TV,* March 11 2022, accessed March 10, 2023. https://www.youtube.com/watch?v=yB8hkZoBQUE

57 "BrahMos Supersonic Missile - Bharat Ne Yeh Missile Ghalti Se Chalaya Ya Jaan Boojh Kar?" - *SAMAA TV,* March 11 2022, accessed March 10, 2023. https://www.youtube.com/watch?v=yB8hkZoBQUE

58 Inam Haque, "What to Make of India's Dangerous 'Error' | *The Express Tribune (T-Magazine),* March 13, 2022, accessed March 10, 2023, https://tribune.com.pk/story/2347631/what-to-make-of-indias-dangerous-error

59 Inam Haque, "India's 'Errant' Missile and Aman Ki Asha," *The Express Tribune,* March 17, 2022, accessed March 10, 2023, https://tribune.com.pk/story/2348281/indias-errant-missile-and-aman-ki-ash

60 "India's Missile Misfire & Pakistan's near Retaliation l Inside the Close Shave," *Mojo Story,* March 16, 2022, accessed March 13, 2023, https://www.youtube.com/watch?v=_pTvqv275ts

61 "Faizan Fakhar," *Islamabad Policy Research Institute,* accessed March 13, 2023, https://ipripak.org/portfolio/faizan-fakhar/

62 Faizan Fakhar, "BrahMos Misadventure: Accidental or Deliberate?," *The Express Tribune,* March 23, 2022, accessed March 13, 2023, https://tribune.com.pk/story/2349154/brahmos-misadventure-accidental-or-deliberate

63 Shahid Alvi, "The Runaway Indian Missile: A Fact Sheet," *Pakistanpolitico,* March 23, 2022, accessed March 18, 2023, https://pakistanpolitico.com/brahmos/

64 "India's 'Accidental' Missile Launch," *Jinnah Institute,* March 25, 2022, accessed March 18, 2023, https://jinnah-institute.org/feature/indias-accidental-missile-launch/

65 "India's 'Accidental' Missile Launch," *Jinnah Institute,* March 25, 2022, accessed March 18, 2023, https://jinnah-institute.org/feature/indias-accidental-missile-launch/
66 "India's 'Accidental' Missile Launch," *Jinnah Institute,* March 25, 2022, accessed March 18 2023. https://jinnah-institute.org/feature/indias-accidental-missile-launch/
67 Ahmed Saeed, "India's Missile 'Mistake' and Global Bias," *T Magazine,* March 27, 2022, accessed March 18, 2023, https://tribune.com.pk/story/2349804/indias-missile-mistake-and-global-bias
68 Haris Bilal Malik, "Brahmos Missile Incident: India's Strategic Irresponsibility or Incompetence," *Strafasia,* March 29, 2022, accessed March 26, 2023, https://strafasia.com/brahmos-missile-incident-indias-strategic-irresponsibility-or-incompetence/
69 Dr Asma Shakir Khawaja, "Brahmos Fire on Pakistan: A Question Mark on Indian Credibility as Responsible State Actor," *Global Village Space*, March 30, 2022, accessed April 4, 2023, https://www.globalvillagespace.com/brahmos-fire-on-pakistan-a-question-mark-on-indian-credibility-as-responsible-state-actor/
70 A. Ahmed, "WEBINAR Organized by Rabita Forum International 'Indian Irresponsible Strategic Behavior: Intentions & Regional& Global Consequences," *Rabita Forum International,* April 1, 2022, accessed April 10, 2023, https://www.rfipakistan.com/webinar-organized-by-rabita-forum-international-indian-irresponsible-strategic-behavior-intentions-regional-global-consequences/
71 GulRukh Nafees, "The Missile Misfire: Time for India and Pakistan to Set down Rules," *Geo News,* April 11, 2022, accessed April 10, 2023, https://www.geo.tv/latest/410798-missile-misfire-time-for-india-pakistan-set-down-rules
72 "Natasha Khan, Author at Centre for Strategic and Contemporary Research," *CSCR,* accessed April 10, 2023, https://cscr.pk/author/natasha-khan/
73 Natasha Khan, "What Does the Indian Missile Misfire Mean for Pakistan?" *Centre for Strategic and Contemporary Research*, April 15, 2022, accessed April 17, 2023, https://cscr.pk/explore/themes/defense-security/what-does-the-indian-missile-misfire-mean-for-pakistan/
74 Muhammad Ali Baig, "Looking Back at the Indian BrahMos Missile Crash in Pakistan," *Global Village Space*, April 19, 2022, accessed April 17, 2023, https://www.globalvillagespace.com/looking-back-at-the-indian-brahmos-missile-crash-in-pakistan/
75 Muhammad Ali Baig, "Strategic Stability-A Gauntlet to India," *Daily Times,* May 20, 2022, accessed April 18, 2023, https://dailytimes.com.pk/937899/strategic-stability-a-gauntlet-to-india/
76 CISS Webinar on "Indian Missile Fiasco: Technical Malfunction and Failure of Diplomacy," *Center for International Strategic Studies,* April 21, 2022, accessed April 27, 2023, https://www.youtube.com/watch?v=LFKKAvJkDoY

Reporter, "Indian Missile Incident Questions Its Arsenal Credibility: Experts," *Dawn,* April 23, 2022, accessed April 27 2023, https://www.dawn.com/news/1686298

CISS Pakistan, "CISS Webinar on 'Indian Missile Fiasco: Technical Malfunction and Failure of Diplomacy,'" *Center for International Strategic Studies,* April 21, 2022, accessed April 27, 2023, https://ciss.org.pk/ciss-webinar-on-indian-missile-fiasco-technical-malfunction-and-failure-of-diplomacy/
77 Sabina Babar, "The Threat of Hypersonic Warfare in Indo-Pacific," *Pakistan Today,* April 22, 2022, accessed May 4, 2023, https://www.pakistantoday.com.pk/2022/04/22/the-threat-of-hypersonic-warfare-in-indo-pacific/
78 Rahat Iqbal, "India's Brahmos Faux Pas and International Response," *Centre for Strategic and Contemporary Research*, May 18, 2022, accessed May 5,

2023, https://cscr.pk/explore/themes/defense-security/indias-brahmos-faux-pas-and-international-response/

79 Samran Ali, "Clearing the Confusion Regarding the Brahmos Missile Fiasco," *Centre for Strategic and Contemporary Research,* May 20, 2022, accessed May 5, 2023, https://cscr.pk/explore/themes/defense-security/clearing-the-confusion-regarding-the-brahmos-missile-fiasco/

80 D. Muhammadi and Rahat Iqbal, "Pakistan's Nuclear Weapons and Deterrence," *Daily Times,* May 31, 2022, accessed May 5, 2023, https://dailytimes.com.pk/943897/pakistans-nuclear-weapons-and-deterrence/

81 Najm Saqib, "Future of Pak-India Relations," *The Nation*, June 6, 2022, accessed May 12, 2023, https://nation.com.pk/2022/06/06/future-of-pak-india-elations/

82 Nadir Ali, "Brahmos, a Failure of Indian Command and Control System," *Pakistan Observer,* June 16 2022, accessed May 12, 2023, https://pakobserver.net/brahmos-a-failure-of-indian-command-and-control-system-by-nadir-ali/

83 Rizwana Abbasi and Mohammed Saeed Uzzaman, "Indian Cruise Missile Misadventure: Malfunction or Malafide Intentions?" *JSSA* 8, no. 1 (2022): 135–44.

84 Ghazala Yasmin Jalil, "India's Accidental Missile Launch--Implications and Challenges Ahead," *Institute of Strategic Studies-Islamabad,* July 6, 2022, accessed May 12, 2023, https://issi.org.pk/wp-content/uploads/2022/07/IB_Misbah_July_28_2022.pdf, p 3–5

85 N. Salik, "Dangers of Inadvertent Escalation in South Asia," in *Nuclear Risks and Arms Control - Problems and Progresses in the Time of Pandemics and War. EAC 2022. Springer Proceedings in Physics*, eds. P. Cotta-Ramusino, M. Lowenthal, L. Maiani, and E. Pellecchia, vol. 291 (Cham: Springer, 2023), https://doi.org/10.1007/978-3-031-29708-3_12

86 Salma Malik, "99 Seconds to Midnight: A Case for Institutionalising Nuclear Risk Reduction Measures in South Asia," *Strategic Studies* 42, no. 1 (2022): 55–59.

87 Sultan M. Hali, "Irresponsible India's Ambitious Missile Program," *Pakistan Observer,* August 28, 2022, accessed May 12, 2023, https://pakobserver.net/irresponsible-indias-ambitious-missile-program-by-sultan-m-hali/

88 Haris Akmal, "South Asia and Nuclear Brinkmanship," *Pakistan Today,* September 1, 2022, accessed May 12, 2023, https://www.pakistantoday.com.pk/2022/09/01/south-asia-and-nuclear-brinkmanship/

89 Keynote Address by Lt. Gen, "Khalid Ahmed Kidwai on 8th CISS-IISS International Workshop," *Center for International Strategic Studies*, September 21, 2022, accessed May 23, 2023, https://www.youtube.com/watch?v=4WHbGmzE1tc Correspondent, "Extremists in Control of Indian Nuclear Arsenal: NCA Adviser," *The Express Tribune*, September 26, 2022, accessed May 23, 2023, https://tribune.com.pk/story/2378687/hindu-extremists-in-control-of-indias-nuclear-arsenal-nca-adviser "CISS–IISS 8th International Workshop on Strategic Stability in South Asia," *Center for International Strategic Studies,* September 21, 2022. Pakistan, accessed May 23, 2023, https://ciss.org.pk/ciss-iiss-8th-international-workshop-on-strategic-stability-in-south-asia/

Baqir Sajjad Syed, "Concern over 'Hindu Extremists' Controlling Indian N-Arsenal," *DAWN.COM*, September 27, 2022, accessed May 23, 2023, https://www.dawn.com/news/1712196

90 Keynote Address by Lt. Gen, "Khalid Ahmed Kidwai on 8th CISS-IISS International Workshop," *Center for International Strategic Studies*, September 21, 2022, accessed May 23, 2023. https://www.youtube.com/watch?v=4WHbGmzE1tc Correspondent, "Extremists in Control of Indian Nuclear Arsenal: NCA Adviser," *The Express Tribune*, September 26, 2022, accessed

May 23, 2023, https://tribune.com.pk/story/2378687/hindu-extremists-in-control-of-indias-nuclear-arsenal-nca-adviser
"CISS–IISS 8th International Workshop on Strategic Stability in South Asia," *Center for International Strategic Studies,* September 21, 2022. Pakistan, accessed May 23, 2023, https://ciss.org.pk/ciss-iiss-8th-international-workshop-on-strategic-stability-in-south-asia/
Baqir Sajjad Syed, "Concern over 'Hindu Extremists' Controlling Indian N-Arsenal," *DAWN.COM*, September 27, 2022, accessed May 23, 2023, https://www.dawn.com/news/1712196

91 Amber Afreen Abid, "Indian Missile Misfire: An Unprecedented Danger," *Global Defense Insight*, April 10, 2022, accessed May 26, 2023, https://defense-talks.com/indian-missile-misfire-an-unprecedented-danger/

92 Moiz Khan, "Urgency of Reviewing India-Pakistan's CBMs & Risk Reduction Measures by Moiz Khan," *CASS,* October 6, 2022, accessed May 26, 2023, https://casstt.com/urgency-of-reviewing-india-pakistans-cbms-risk-reduction-measures/

93 Shahid Hameed, "India's Regional Instability Scam," *The Nation,* December 3, 2022, accessed May 26, 2023, https://www.nation.com.pk/03-Dec-2022/india-s-regional-instability-scam

94 Syed Qamar Afzal Rizvi, "India Must Discern the Perils of the Missile Incident," *Pakistan Observer,* December 15, 2022, accessed May 26, 2023, https://pakobserver.net/india-must-discern-the-perils-of-missile-incident-by-syed-qamar-afzal-rizvi/

4

GLOBAL MEDIA COVERAGE

The overall aim of the chapter is to consolidate the key themes from the global media discourse on the missile accident into a coherent narrative. The global print and electronic media's coverage of India's accidental missile launch was constrained by various factors, even as it conveyed strategically critical insights on the missile accident. Perhaps the most important reason for the subdued coverage was that the incident did not result in any casualties. The coverage by the global media would have almost certainly been more intense, had the missile hit a commercial airliner or caused deaths and injuries by crashing in a thickly populated urban area eliciting a stronger response from Pakistan.

Western media outlets in particular were pre-occupied with the Russian invasion of Ukraine. Sustained media coverage of the missile accident was also constrained by the secrecy surrounding India's Court of Inquiry which resulted in the available information reducing to a trickle between early April (when the preliminary results of the inquiry were reported by the Indian media) and late August 2022 (when the inquiry's final results were made public). Indian reticence in sharing information also resulted in the global media reporting more on statements and developments from the Pakistani side.[1]

In a dynamic similar to that of the Indian and Pakistani media, the international media also became a conduit for the sentiments of the Indian and Pakistani security establishments in the form of anonymous statements by either senior officials or people close to the establishment. Such calculated leaks resulted in strategically crucial revelations being brought out by the international media including Pakistani plans for a retaliatory strike to the Indian errant missile which was later aborted and the Indian move to

DOI: 10.4324/9781003543909-5

shutdown the missile system in the immediate aftermath of the misfire to prevent further launches.[2]

The international media also relied heavily on the information provided by news agencies such as *Reuters,* resulting in limited independent analysis of the missile accident. For the purposes of this chapter, the global media will be understood to mean print and electronic media with base of operations outside India and Pakistan (although articles by Indo-Canadian outlets such as *EurAsian Times* have also been included).

Key themes in the global media coverage such as its surprise over the relative calm in the aftermath of the incident, it's reporting of the Pakistani preparations and eventual abandonment of a retaliatory strike, Pakistan's demand for a joint probe and India's announcement to review its standard operating procedures are covered. The chapter also summarizes other themes such as the doubts raised over India's lack of security protocols and technical safeguards, Pakistan's questioning of India's command and control system, praise for Pakistan's response of not deploying a retaliatory strike by Indian journalists quoted by the international media, worries over a similar Indian accidental launch into China and articles on results of India's Court of Inquiry identifying human error as the cause of the accident.

Pakistani Claim of a Missile Fired from Sirsa, India's Claim of a Technical Malfunction

The global media's initial reaction to the missile accident was one of reporting the facts. *Reuters* reported on 11 March the Indian Defence Ministry's claim that India had accidentally fired a missile into Pakistan due to a "technical malfunction." It also reported details provided by the Pakistani military, including the claim that the missile was fired from the northern Indian city of Sirsa.

> A Pakistani military spokesman told a news conference on Thursday evening that a "high-speed flying object" originating from the northern Indian city of Sirsa had crashed in eastern Pakistan. The flight path of this object endangered many national and international passenger flights both in Indian and Pakistani airspace as well as human life and property on ground," he said. A Pakistan air force official said the object, flying at 12,000 metres (40,000 feet) and three times the speed of sound, had flown 124km (78 miles) into Pakistani airspace.[3]

An article on the same day by Ayaz Gul in *Voice of America* reported the claim by the Pakistani military of an unarmed supersonic missile crashing into its territory and the somewhat contradictory claim of Pakistani Federal Minister Asad Umar that the Pakistan Air Force had shot down the Indian missile. The article also focused on the Pakistani National Security Adviser

Moeed Yusuf's strident criticism of the delay in India's statement, call for greater involvement of the international community and the need for India to investigate whether the missile launch was accidental or deliberate.[4]

Bloomberg reported the Indian admission that it had accidentally fired a missile into Pakistan during routine maintenance and the Pakistani protest that the missile's trajectory had endangered other civilian planes.[5] Shan Li and Vibhuti Agarwal adopted a neutral tone in the *Wall Street Journal* and reported that India had blamed a technical malfunction for the accidental "mishap" resulting in the launch of the missile into Pakistan which caused minor damage to civilian property but did not result in any casualties.[6] They did not speculate on the type of missile fired.

An article by *Reuters Islamabad* published in *The Guardian* quoted major portions of the Indian Ministry of Defence's terse initial statement blaming the accident on a technical malfunction and its announcement of a high-level Court of Inquiry. The article also quoted an anonymous senior Pakistani security official expressing strong reservations over India's "non-chalant" admission of the missile misfire and contending that the released missile was most probably a BrahMos supersonic cruise missile.

> One senior Pakistani security official told Reuters, on condition of anonymity, that the incident could have escalated into a "critical untoward situation". "The admission that it was a missile was very nonchalant," he said. "What does this say about their safety mechanisms and the technical prowess of very dangerous weapons? The international community needs to have a very close look at this." The official said it was possibly a BrahMos missile – a nuclear-capable, land-attack cruise missile jointly developed by Russia and India.[7]

The Adversarial India-Pakistan Relationship, Indian Missiles in Ready to Launch Positions

A *Deutsche Welle* (German) article also reproduced major portions of the Indian Defence Ministry's statement and the Pakistani complaints regarding the accidental firing showing India's safety mechanisms and technological capabilities in poor light while also calling for the international community to take note. The article also situated the accidental firing in the context of the airstrikes carried out by India and the retaliatory strikes by Pakistan in the aftermath of the 2019 Balakot crisis and then warned about the escalatory potential of such accidents.

> India and Pakistan both have nuclear weapons, and military experts have repeatedly warned of the risk of miscalculation or nuclear accidents. The two countries veered dangerously close to a full-out war in 2019, when

> both launched airstrikes across the Line of Control, the de facto border between India and Pakistan.[8]

In a *Forbes* magazine article, staff writer Carlie Porterfield also chose to contextualize the accident against the historically tense and adversarial character of the India-Pakistan relationship.

> Pakistan's Ministry of Foreign Affairs said in a statement it summoned India's *chargé d'affaires*—a diplomat who takes the place of an absent ambassador—to Islamabad to protest the "unprovoked violation" of Pakistan's airspace....Since Pakistan and India both gained independence from British colonial rule in 1947, the countries have fought in three wars against each other, and tension remains high over the disputed Kashmir territory in the Himalayas, which both nations claim as their own. The dispute most recently peaked in February 2019, when an Indian fighter plane was shot down over the Pakistan-run area of Kashmir and its pilot was captured by the Pakistani air force.[9]

An 11 March article on the *Reuters* website channeled the worries of an anonymous senior Pakistani official in the aftermath of the missile accident as to whether Indian missiles were always kept in "ready-to-launch" positions pointed at Pakistan without safeguards.

> According to the U.S.-based Arms Control Association, the missile's range is between 300 kilometres and 500 kilometres, making it capable of hitting Islamabad from a northern Indian launch pad. The (anonymous senior) Pakistani official (speaking to *Reuters*) wondered if the incident meant that India had "missiles in ready-to-launch positions and pointed at Pakistan, and that too without any safeguard of a command and control system.[10]

Both *CBC News* and *Taipei Times* reproduced the aforementioned *Reuters* dispatch in its entirety.[11]

Surprise over the post-accident Calm, Call for Joint Probe, Risk of India-China Accident

The initial focus of the international media was mostly on reporting the developing events surrounding the incident which later shifted to analysis of its aftermath and its implications. Reporting for the *New York Times* on 12 March, Mujib Mashal and Salman Masood expressed surprise over the relative calm that ensued following India's accidental launch.

> One nuclear-armed state fired a cruise missile at another nuclear-armed state this week. They were not at war, and it did not start one. On Friday, India acknowledged that one of its missiles had mistakenly been fired into Pakistan two days earlier. Pakistan criticized India's "callousness and ineptitude" in a "nuclear environment." And that, so far, has been the end of the matter—a subdued aftermath that many saw as nothing short of a small miracle....Analysts in India commended the Pakistani military, the country's most powerful institution, for its reserved response to the missile firing, which apparently caused no casualties. That muted reaction seems to have headed off what could have become a disastrous escalation....But the episode is bound to raise concerns about the safety of India's weapons systems, and about the government's credibility on that subject. India waited 48 hours to confirm that the accident had happened, and Pakistani officials said they had received no information about it from their Indian counterparts in the meantime.[12]

Al Jazeera focused on Pakistan's rejection of India's Court of Inquiry and its demand for a joint probe into the incident.

> Pakistan says it has taken note of India's statement on the "accidental firing" of a missile into its territory, demanding a joint investigation to "accurately establish the facts surrounding the incident". The grave nature of the incident raises several fundamental questions regarding security protocols and technical safeguards against accidental or unauthorised launch of missiles in a nuclearised environment," the foreign ministry said in a statement on Saturday, adding that India must provide a clearer picture of the circumstances that led to the missile's firing.[13]

Writing for the *EurAsian Times*, Ashish Dangwal reproduced the seven "vital questions" posed by the Pakistan Foreign Office to India in the immediate aftermath of the missile firing. He also focused on the bafflement of unidentified defense experts over the chain of causality that caused the accidental launch despite the presence of multiple safeguards.

> Military experts are baffled by the occurrence since launching a missile system necessitates a preparation phase, the identification of a target, and the toggling of many switches. Also, there seems to be no BrahMos base in Sirsa, the missile system may be relocated from one area to another using specialized vehicles. The Andaman and Nicobar Command and the eastern flank are mainly used to test missiles like the BrahMos.[14]

A 13 March article in *The Global Times,* China's nationalist newspaper by defense reporter Liu Xuanzun expressed concerns regarding the efficacy of

Indian security protocols and technical standards. He also worried that a similar unintentional accident on the Indo-China border could derail the attempts by both countries to resolve their ongoing border dispute.

> After an Indian supersonic missile was allegedly accidentally fired into Pakistani territory last week, Pakistan demanded a joint probe on Saturday, as Pakistan said it doubts India's internal inquiry will answer its questions over India's capabilities to handle such weapons properly. Other countries neighboring India including China should also be concerned about India's lack of effective security protocols or technical safeguards, as this could potentially lead to unexpected consequences under complex situations, analysts said on Sunday....China and India just concluded the 15th round of commander level meeting on Friday, as the two sides reached consensus on agreeing to maintain dialogue via military and diplomatic channels to reach a mutually acceptable resolution to the remaining issues at the earliest time, continuing the positive atmosphere since the 14th round of talks. But an unintentional accident caused by India's incapability to handle its weapons could seriously complicate the situation, experts said.[15]

India's Dented Image, China Recommends Establishing Reporting Mechanisms

Umair Jamal, a Lahore-based correspondent for *The Diplomat*, a US-based publication analyzed that the accident had dented India's image as a responsible nuclear power and called for a joint probe to ascertain all the facts surrounding the incident. The article erroneously identified the BrahMos cruise missile as a ballistic missile.

> Last week, India accidentally fired a ballistic missile at its nuclear armed arch-rival, Pakistan. Islamabad's measured response to the event, which has been lauded widely, perhaps averted a calamity. However, the event, and India's delayed and casual response to it, has not only set alarm bells ringing in Islamabad but internationally as well, denting New Delhi's "credibility as a responsible, mature and accountable nuclear-weapon state."...In any case, for India and Pakistan to develop a better and more comprehensive protocols to prevent such incidents, New Delhi should coordinate with Islamabad to ensure that all important questions regarding the event are satisfactorily answered.[16]

Rimal Farrukh, South Asia correspondent for *VICE World News* warned of the potential dangers of missile accidents due to human error by highlighting the death and injuries caused by the Titan 2 missile accident in 1980.

> Human error has caused nuclear accidents in the past, including in the world's dominant nuclear power, the United States. In 1980, a Titan 2 missile equipped with a nuclear warhead exploded in Damascus, Arkansas when a maintenance worker accidentally dropped a wrench socket into a shaft and pierced the missile, releasing explosive fuel. The blast killed one person and injured 21 others.[17]

A 15 March article in *Al Jazeera* covered India's ongoing review of standard operating procedures in the aftermath of the accidental launch into Pakistan. The article also informed about the Pakistani and Chinese responses to the Indian announcement of a high-level inquiry into the accident.

> India is conducting a review of its procedures for operations, maintenance and inspection of weapons systems after accidentally launching a missile into Pakistan last week, its defence minister says...Pakistan has asked for a joint probe into the incident, demanding clarifications from New Delhi over its safety mechanism to prevent accidental missile launches. Pakistan's foreign minister raised the issue with United Nations Secretary-General Antonio Guterres by telephone on Monday, where he stressed the need for the international community, including the Security Council, to address the situation. China asked the South Asian neighbours to set up a reporting mechanism to avoid such incidents and share information, said Zhao Lijian, a spokesman of its foreign ministry.[18]

Pakistan's Aborted Retaliatory Strike, Missile Launched from Ambala, Hotline Not Used

Writing for *Bloomberg.com*, Sudhi Ranjan Sen and Faseeh Mangi quoted unnamed "people familiar with the matter" to inform that the "errant Indian missile" nearly led to a Pakistani retaliatory strike. The article also relied on unnamed Indian sources to claim that the errant missile had been launched from the garrison town of Ambala (and not Sirsa as claimed by Pakistan) and that the accident was a result of human and technical errors during a routine exercise.

> An accidental missile fired by India last week prompted Pakistan to prepare a retaliatory strike, people familiar with the matter said, showing how close the nuclear-armed neighbors came to blows over a potentially disastrous mistake. Pakistan had prepared to launch a similar missile to strike India but held back because an initial assessment indicated something was amiss, people familiar with the matter said....The Indian Air Force fired the BrahMos medium-range cruise missile on March 9 from the garrison town of Ambala, about 200 kilometers (125 miles) north of the capital New Delhi, according to people familiar with the matter in

> India. The mishap occurred due to human and technical errors during a routine exercise to check systems capable of taking offensive action in war, the people said. Still, after the launch India didn't use the direct hotline between the top army commanders on both sides to inform Pakistan, the people said, asking not to be identified due to the sensitivity of the information. Instead, Air Force officials moved to shut down the missile systems to avoid any further launches, they said....India's nuclear warheads and delivery mechanisms are always kept separate in peace time. The country's nuclear-armed missiles are under the Strategic Forces Command headed by the prime minister, and several layers of permissions are required before they can be even moved.[19]

An identical version of the *Bloomberg* article was carried by the Hong-Kong-based *South China Morning Post* on 16 March.[20] A 15 March article on *BBC.com* also covered India's investigation into the causes of the missile accident which was "accidentally released during routine maintenance and inspection of a missile unit."[21] Writing for the *Voice of America*, Ayaz Gul informed readers about the ongoing Indian review of standard operating procedures after the missile misfire into Pakistan, the Chinese and American reactions to the accident, the Pakistani appreciation of China's statement and its disappointment with the US stance on the matter.

> On Monday, China urged India and Pakistan to establish a reporting mechanism for sharing information to avoid such incidents in the future. "Pakistan and India are both important countries in South Asia, bearing responsibilities for maintaining regional security and stability," Foreign Ministry spokesman Zhao Lijian told reporters in Beijing. The U.S. State Department spokesperson also was asked at a Monday news conference for Washington's response to India's accidental firing of the missile into Pakistan. "Well, we have no indication, as you also heard from our Indian partners, that this incident was anything other than an accident... They [India] issued a statement...to explain precisely what had happened. We don't have a comment beyond that," Ned Price said. While (Pakistan's Foreign Minister) Qureshi said he appreciated China's statement, he criticized the U.S. spokesperson for not taking a "clear position" on the issue. "He [Price] understands its importance, but I am sorry to say he skirted the issue," the foreign minister said.[22]

A 15 March article in the *Daily Sabah*, a Turkish pro-government newspaper in partnership with *Reuters* also covered India's "thorough inspection" of standard operating procedures for weapons systems after the accidental missile launch. The paper also informed that Pakistan had demanded an explanation "regarding measures and procedures in place to prevent the accidental missile launch, type and specification of the missile, flight path and India's failure to inform them about the incident."[23]

A Danger to Domestic and International Commercial Flights—Preliminary Findings of Indian Probe

A 17 March article by *Bloomberg*, carried by the Riyadh-based *Al Arabiya News* explained how the accidental missile could have hit several domestic and international commercial jets that were operating in the vicinity around the same time:

> A missile that was accidentally fired from India last week narrowly missed dozens of commercial jets that were in the same skies. Several planes passed through the direct trajectory of the missile that day, which flew from the Indian garrison town of Ambala and ended up in Mian Channu in Eastern Pakistan. They included a Flydubai jet heading to Dubai from Sialkot, an IndiGo plane from Srinagar to Mumbai and an Airblue Ltd. flight from Lahore to Riyadh. All crossed the missile's trajectory within an hour of its accidental launch, data from flight-tracking application Flightradar24 show. Other international flights in the vicinity of the missile's trajectory—and within its range—included a Kuwait Airways Co. jet heading to Guangzhou from Kuwait City, a Saudi Arabian Airlines flight to Riyadh from New Delhi, and a Qatar Airways service from Kathmandu to Doha, the data show. No advisory to pilots operating in the vicinity—known as a notice to airmen or NOTAM—was issued.[24]

A 24 March article by New-Delhi based journalist Taniya Dutta in *The National,* a Abu Dhabi (UAE) based English language newspaper informed that India had successfully test fired a "nuclear-capable" BrahMos missile in the days following the missile accident.

> India has successfully test-fired the nuclear-capable surface-to-surface BrahMos supersonic cruise missile from the remote Andaman and Nicobar Islands. India's Defence Research and Development Organisation said the extended range missile was test-fired from a land-based launcher from an island in the eastern Indian Ocean. BrahMos, a nuclear-capable missile, is one of the world's fastest cruise missiles and can be fired from land, sea and air. It came under the spotlight this month after New Delhi said it "accidentally" fired a missile into Pakistan....The launch was witnessed by air chief marshal VR Chaudhari and other defence officials, authorities said. They said the missile hit its target with pinpoint accuracy.[25]

After a brief lull, the international media coverage of the missile accident picked up again with a 10 April article by China's Xinhua News Agency reporting on the preliminary findings of India's Court of Inquiry.

> The Indian Air Force (IAF) has blamed officials from its missile squadron for an incident which accidentally launched a missile into Pakistan....The inquiry was done by a senior IAF official. "More than one official has been found blameworthy for the incident which was totally avoidable. The guilty officials would be given a swift and severe punishment," the Asian News International (ANI) quoted government sources as saying. The court of inquiry also found a clear violation of standard operating procedures by the officials concerned, they said.[26]

Missile Accident and Export Potential of BrahMos, India's Weak Command and Control

The Indo-Canadian *EurAsian Times*, carried an article by Indian defense journalist Sakshi Tiwari on 8 April which covered the impact of the missile misfire on the export potential of the BrahMos to the Philippines and several other interested countries.

> India's aim to rake in $5 billion through defense export by 2025, with the BrahMos supersonic cruise missile high on the list may see a setback. The accidental firing of the missile into Pakistan and crippling sanctions on Russia over its invasion of Ukraine could potentially hinder India's ambitious plan for its exports....In a debut deal for BrahMos missiles, India secured a $375 million contract from the Philippines to supply three BrahMos ground-based anti-ship missile systems....Being the first buyer of the missile, Manila has reportedly been keeping a close eye on the incident....Shambhu Kumaran, India's envoy to Philippines, told the media that there were some questions from Manila about the incident, and he had spoken with Defense Secretary Delfin Negrillo Lorenzana about it...In addition to the Philippines, Vietnam and the United Arab Emirates (UAE) are said to be interested in purchasing the BrahMos... India, according to sources, has held similar talks with Argentina, Brazil, Indonesia, and South Africa.[27]

Eurasia Review, an independent publication based in Oregon in the US carried an opinion piece on 27 June by Amber Afreen Abid, a Research Officer at the Islamabad-based Strategic Vision Institute (SVI) worrying about the impacts of Indian supersonic and hypersonic weapons for India's nuclear doctrine and South Asian security. The article also accused the Indian political elite of being "immature."

> Recently, Atul Rane, CEO and MD, BrahMos Aerospace said that in five to six years, India will be able to have the first hypersonic missile.... Owing to the volatile situation in South Asia, with the absence of any conflict resolution treaties and agreements, the innovation in technology in South Asia leads to the change in the nuclear doctrines as well. Pakistan maintains a policy of minimum credible deterrence, but that minimum is directly proportional to the advancements made by the adversary in offensive technology and ultimately in the nuclear doctrine...The recent BrahMos misfire incident into the Pakistan territory indicates the weak command and control structure of India...Pakistan has always made efforts for restoring regional peace and stability, which India has always tried to destabilize due to its immature ruling authority....Moreover, the world has seen numerous instances of Uranium theft in India, which indicates weak safety and security protocols and weak Command and Control structure in India to handle such precarious technology.[28]

Result of India's Court of Inquiry and Pakistan's Rejection

There followed another lull with media coverage picking up after the final result of the inquiry became known. A 23 August article in *Reuters* highlighted the Indian government's decision to terminate the services of three IAF officers and praised India and Pakistan for their restrained handling of the missile accident. The article also informed that the errant BrahMos missile was nuclear-capable.

> The Indian Air Force said on Tuesday the government had sacked three officers for accidentally firing a missile into Pakistan in March, an incident that the two nuclear-armed rivals handled calmly as there were no casualties. The BrahMos missile – a nuclear-capable, land-attack cruise missile jointly developed by Russia and India - was fired on March 9, prompting Pakistan to seek answers from New Delhi on the safety mechanisms in place to prevent accidental launches.[29]

The *BBC* also reported on the findings of the Indian probe, the subsequent sacking of the officers held responsible and the key conclusion of the investigation that the accident occurred due to deviation from Standard Operating Procedure (SOPs).[30] The *South China Morning Post*, re-published an article by *Agence France-Presse* on India's sacking of three officers for the missile accident. India's 48-hour delay in acknowledging the accidental launch of the missile with its terse 11 March statement was by now forgotten with the article crediting India for immediately acknowledging the accidental release. The article also praised India and Pakistan for their calm handling of the incident despite their historically hostile relationship. The BrahMos missile was also identified as India's frontline nuclear weapon delivery platform:

> Islamabad had said that a "supersonic flying object" cruising at 12km (40,000 feet) from India had violated its airspace and hit an uninhabited civilian area on March 9. Delhi had immediately acknowledged the mistake, which Islamabad said endangered civilian planes and lodged a diplomatic protest...But in a rare instance, both the nuclear-armed nations unusually downplayed the incident despite a history of deep mistrust and hostility. BrahMos is one of India's frontline nuclear-armed projectiles and is believed to be among the fastest cruise missiles in the world.[31]

In the immediate aftermath of India's announcement of its probe's findings, a 24 August article in *Al Jazeera* reported on Pakistan's initial reaction rejecting India's findings and reiterating its demand for a joint probe.

> Pakistan has rejected India's closure of the incident of the firing of a supersonic missile into Pakistani territory on March 9, and reiterated a demand for a joint probe, the Ministry of Foreign Affairs said on Wednesday. "India has not only failed to respond to Pakistan's demand for a joint inquiry but has also evaded the questions raised by Pakistan regarding the command-and-control system in place in India, the safety and security protocols and the reason for India's delayed admission of the missile launch."[32]

CNN and *ABC News* also carried articles informing about the Pakistani rejection of the Indian probe's findings and its calls for a joint probe.[33]

Analysis

The international media's coverage of the missile accident was sparse and lacked strong editorial viewpoints that would have normally been carried on an issue as serious as the accidental release of a cruise missile by a nuclear armed state into another which considers it to be nuclear-capable. The aforementioned void was in stark contrast to the regular coverage and strong editorial stances taken by the Indian and Pakistani media. There were multiple reasons for this including the pre-occupation of the international (especially western) media with the war in Ukraine, the errant missile being unarmed, not causing any loss of life (either on ground or in the air by slamming into a commercial airline) and not crashing into a strategic facility such as a defense base or nuclear installation. The accident's occurrence during a period of relative calm in the India-Pakistan relationship and Pakistan's decision to not retaliate to the intruding missile also played a role in reducing the salience of the issue for the global media.

As a result, the coverage of the event was subdued and restricted to mostly reporting Pakistani claims and Indian counterclaims and situating the missile accident in the adversarial context of India-Pakistan relations while only occasionally relaying revealing insights through anonymous statements by Indian and Pakistani officials. *Bloomberg*'s relaying of the claim by anonymous Pakistani sources of Pakistan's brief consideration and eventual aborting of a retaliatory strike was considered important enough to be reported widely in the Indian and Pakistani press. Also revealing was *Bloomberg*'s carrying of an anonymous statement from Indian officials that India did not use the hotline and instead the Indian Air Force moved to quickly shutdown the missile systems after the accident to prevent further launches. Another key snippet relayed by the international media was an anonymous statement by Pakistani officials worrying about whether Indian missiles were always pointed at Pakistan and always maintained in launch-ready mode. Coverage also focused on the Pakistani claim of the missile being launched from Sirsa in the Indian state of Haryana and the Indian claim of the missile actually being launched from the garrison city of Ambala. The refusal of the US to criticize India for either the accidental launch or the delay in acknowledging it and the Chinese statement calling for the establishment of "reporting mechanisms" between India and Pakistan was also reported. Coverage was also given to viewpoints about the escalatory potential of an accidental Indian missile launch into China.

The surprise expressed by sections of the global media regarding the manner in which both India and Pakistan successfully managed to avoid an escalation following the missile accident points to the power of the deeply embedded narrative of India-Pakistan hostility which leads to expectations that any transgression will unavoidably lead to rapid escalation. Such a narrative demonstrates a lack of understanding of the particularities of the India-Pakistan relationship and of conflict dynamics in South Asia. The international media's intermittent interest in the missile accident also led it to miss the evolution of the Indian government's stated rationale from the technical malfunction to human error.

And finally, the coverage by the international media of the missile accident was similar to the coverage of the Indian and Pakistani media in terms of highlighted issues such as the Indian missile being a danger to domestic and international commercial flights, the export potential of the BrahMos, preliminary findings of the Court of Inquiry and the rejection of its final conclusion by Pakistan.

Notes

1 Mujib Mashal and Salman Masood, "India Accidentally Fires a Missile at Pakistan. Calm Ensues," *The New York Times,* March 12, 2022, sec. World, accessed August 5, 2023, https://www.nytimes.com/2022/03/12/world/asia/india-pakistan-missile.html

2 Sudhi Sen and Faseeh Mangi. "Errant Indian Missile Nearly Led to Pakistan Retaliatory Strike," *Bloomberg.com*, March 15, 2022, accessed August 6, 2023, https://www.bloomberg.com/news/articles/2022-03-15/errant-indian-missile-nearly-led-to-pakistan-retaliatory-strike

3 Reuters, "India Fires Missile into Pakistan by Accident," *South China Morning Post,* March 11, 2022, accessed August 6, 2023, https://www.scmp.com/news/asia/east-asia/article/3170205/india-fired-missile-pakistan-accident-defence-ministry-said.
Reuters, "India Says It Accidentally Fired Missile into Pakistan," *CNBC*, March 11, 2022, accessed August 6, 2023, https://www.cnbc.com/2022/03/11/india-says-it-accidentally-fired-missile-into-pakistan.html

4 Ayaz Gul, "India Confirms It Accidentally Fired Missile into Pakistan," *VOA*, March 11, 2022, accessed August 7 2023, https://www.voanews.com/a/india-confirms-it-accidentally-fired-missile-into-pakistan-/6480872.html

5 Muneeza Naqvi and Sudhi Ranjan Sen, "India Says It 'Accidentally' Fired a Missile Into Pakistan," *Bloomberg.com*. March 11, 2022, accessed August 7, 2023, https://www.bloomberg.com/news/articles/2022-03-11/pakistan-says-unarmed-missile-from-india-landed-in-its-territory

6 Shan Li and Vibhuti Agarwal, "India Accidentally Fires Missile into Pakistan," *Wall Street Journal,* March 11, 2022, sec. World, accessed August 7, 2023, https://www.wsj.com/articles/india-accidentally-fires-missile-into-pakistan-11647023894

7 Reuters, "India Fires Missile into Pakistan after 'Technical Malfunction,'" *The Guardian,* March 11, 2022, sec. World news, accessed August 7, 2023, https://www.theguardian.com/world/2022/mar/11/india-fires-missile-into-pakistan-after-technical-malfunction
Reuters, "India Says It Accidentally Fired a Missile into Pakistan," *CNN*, March 12, 2022, accessed August 7, 2023, https://www.cnn.com/2022/03/11/asia/india-pakistan-missile-intl-hnk/index.html

8 Reuters, "India Reports 'accidental Firing' of Missile into Pakistan," *DW*, March 11, 2022, accessed August 9, 2023, https://www.dw.com/en/india-says-it-accidentally-launched-a-missile-into-pakistan/a-61100492

9 Carlie Porterfield, "India Confirms It Accidentally Fired A Missile Into Pakistan," *Forbes*, March 11, 2022, accessed August 9, 2023, https://www.forbes.com/sites/carlieporterfield/2022/03/11/india-confirms-it-accidentally-fired-a-missile-into-pakistan/

10 Asif Shahzad, Krishna N. Das, Gibran Naiyyar Peshimam, Asif Shahzad, and Krishna N. Das, "India Says It Accidentally Fired Missile into Pakistan," *Reuters,* March 11, 2022, sec. Asia Pacific, accessed August 9, 2023, https://www.reuters.com/world/asia-pacific/pakistan-seeks-answers-india-after-crash-mystery-flying-object-2022-03-10/

11 Reuters, "Accidental Missile Launch into Pakistan Caused by 'technical malfunction,' India Says," *CBC News,* March 11, 2022, accessed August 9, 2023, https://www.cbc.ca/news/world/india-pakistan-missile-launch-technical-malfunction-1.6381822

Reuters, "India Admits to 'Accidental' Missile Fire," *Taipei Times,* March 13, 2022, accessed August 10, 2023, https://www.taipeitimes.com/News/world/archives/2022/03/13/2003774702

12 Mujib Mashal and Salman Masood, "India Accidentally Fires a Missile at Pakistan. Calm Ensues," *The New York Times*, March 12, 2022, accessed August 10, 2023, https://www.nytimes.com/2022/03/12/world/asia/india-pakistan-missile.html

13 Al Jazeera and News Agencies, "Pakistan Demands Joint Probe into 'Accidental' India Missile Fire," *Al Jazeera* March 12, 2022, accessed August 11, 2023, https://www.aljazeera.com/news/2022/3/12/pakistan-demands-joint-probe-after-accidental-india-missile

14 Ashish Dangwal, "Who Was Handling India's Supersonic Missile? Pakistan Asks 7 Vital Questions To New Delhi Over Accidental Missile Firing," *The EurAsian Times*, March 12, 2022, accessed August 11, 2023, https://www.eurasiantimes.com/who-was-handling-indias-supersonic-missiles-pakistan-asks-7-vital-questions/

15 Liu Xuanzun, "India's 'Accidental' Missile Fire into Pakistan Raises Questions, Concerns," *Global Times,* March 13, 2022, accessed August 11, 2023, https://www.globaltimes.cn/page/202203/1254749.shtml
Liu Xuanzun, "India's 'Accidental' Missile Fire into Pakistan Raises Questions, Concerns," *People's Daily Online*, March 14, 2022, accessed August 12, 2023, http://en.people.cn/n3/2022/0314/c90000-9970763.html

16 Umair Jamal, "Pakistan Alarmed Over India's Accidental Missile Firing," March 14, 2022, accessed August 12, 2023, https://thediplomat.com/2022/03/pakistan-alarmed-over-indias-accidental-missile-firing/

17 Rimal Farrukh, "Nuclear-Armed India 'Accidentally' Fired a Missile Into Nuclear-Armed Pakistan," *Vice*, March 14, 2022, accessed August 12, 2023, https://www.vice.com/en/article/n7n5wx/india-accidentally-fired-missile-pakistan

18 News Agencies, "India Reviews Ops after Missile Accidentally Fired into Pakistan," March 15, 2022, accessed August 13, 2023, https://www.aljazeera.com/news/2022/3/15/india-reviews-procedures-missile-accidentally-fired-pakistan

19 Sudhi Sen and Faseeh Mangi, "Errant Indian Missile Nearly Led to Pakistan Retaliatory Strike," *Bloomberg.com,* March 15, 2022, accessed August 13, 2023, https://www.bloomberg.com/news/articles/2022-03-15/errant-indian-missile-nearly-led-to-pakistan-retaliatory-strike

20 Bloomberg, "Pakistan Had Strike Plan after India's Accidental Missile Firing," *South China Morning Post,* March 16, 2022, accessed August 13, 2023, https://www.scmp.com/news/asia/south-asia/article/3170655/pakistan-had-retaliatory-strike-plan-after-indias-accidental

21 BBC, "India Investigates Missile That Landed in Pakistan," *BBC News*, March 15, 2022, accessed August 13, 2023, https://www.bbc.com/news/world-asia-india-60747342

22 Ayaz Gul, "India Says Reviewing Policy After Misfiring of Missile Into Pakistan," *VOA*, March 15, 2022, accessed August 16, 2023, https://www.voanews.com/a/india-says-reviewing-policy-after-misfiring-of-missile-into-pakistan/6486346.html

23 Reuters with Daily Sabah, "India Reviews Procedures after Accidental Pakistan Missile Launch," *Daily Sabah,* March 15, 2022, accessed August 16, 2023, https://www.dailysabah.com/world/asia-pacific/india-reviews-procedures-after-accidental-pakistan-missile-launch

24 Bloomberg, "India's Stray Missile That Fell in Pakistan Put Dozens of Passenger Jets At Risk," *AlArabia News*, March 17, 2022, accessed August 18, 2023,

https://english.alarabiya.net/News/world/2022/03/17/India-s-stray-missile-that-fell-in-Pakistan-put-dozens-of-passenger-jets-at-risk

25 Taniya Dutta, "India Test-Fires Nuclear-Capable BrahMos Supersonic Missile," *The National*, March 24, 2022, accessed August 18, 2023, https://www.thenationalnews.com/world/2022/03/24/india-test-fires-nuclear-capable-brahmos-supersonic-missile/

26 Xia Hua, "Indian Air Force Blames Officials for Missile That Accidentally Landed in Pakistan," *Xinhua*, April 10 2022, accessed August 18, 2023, https://english.news.cn/asiapacific/20220411/fc27f288eb7045d7bfc8b264b8493733/c.html

27 Sakshi Tiwari, "BrahMos Missile: India's 'Most Promising' Weapon of Export Hits Turbulence Over Russia Sanctions, Pakistan Incident?" *The EurAsian Times*, April 8, 2022, accessed August 19, 2023, https://www.eurasiantimes.com/sanctions-on-russia-accidental-firing-of-brahmos-missile/

28 Amber Afreen Abid, "Indian Nuclear Missile Proliferation: Effect on South Asian Strategic Stability," *Eurasia Review*, June 27, 2022, accessed August 20, 2023, https://www.eurasiareview.com/27062022-indian-nuclear-missile-proliferation-effect-on-south-asian-strategic-stability-oped-2/

29 Krishna Das, "India Fires Three Officers for Accidentally Launching Missile into Pakistan in March," *Reuters*, August 23, 2022, accessed August 20, 2023. India, https://www.reuters.com/world/india/india-fires-three-officers-accidentally-firing-missile-into-pakistan-march-2022-08-23/

30 BBC News, "India Sacks Officers for Accidentally Firing Missile into Pakistan," August 24, 2022, accessed August 21, 2023, https://www.bbc.com/news/world-asia-india-62655736

31 Agence France Press, "India Sacks Officers for Accidentally Firing Missile into Pakistan," *South China Morning Post*, August 24, 2022, accessed August 21, 2023, https://www.scmp.com/news/asia/south-asia/article/3189947/india-sacks-officers-accidentally-firing-missile-pakistan

The Associated Press, "India Fires 3 Military Officers for Pakistan Missile Misfire," *Toronto Star*, August 23, 2022, accessed August 21, 2023, https://www.thestar.com/news/world/asia/india-fires-3-military-officers-for-pakistan-missile-misfire/article_723c6c10-ac28-5693-836a-1fe8c3eece66.html

32 "Pakistan Rejects India's Closure of Missile Firing Incident," *Al Jazeera*, August 24, 2022, accessed August 21, 2023, https://www.aljazeera.com/news/2022/8/24/pakistan-rejects-indias-closure-of-missile-firing-incident.

33 Reuters, "Pakistan Rejects India's Closure of Missile Firing Incident into Its Territory," *CNN*, August 25, 2022, accessed August 21, 2023, https://www.cnn.com/2022/08/24/asia/pakistan-india-missile-accident-intl-hnk-ml/index.html

Associated Press, "India Dismisses Military Officers for Pakistan Missile Misfire," *ABC News*, August 23, 2022, accessed August 22, 2023, https://www.abc.net.au/news/2022-08-24/india-fires-three-military-officers-for-pakistan-missile-misfire/101363778

5

INTERNATIONAL EXPERT ANALYSES

The implications of the accidental missile launch from India into Pakistan were discussed and debated by strategic experts in the US, China and other countries. The experts analyzed both the specifics of the incident and its implications for South Asian and global security. Important issues were discussed including the implications of the missile accident for India's international image as a responsible state,[1] the unusual accidental launch of the missile during routine maintenance,[2] the need to extend the 2005 ballistic missile test pre-notification agreement to also include cruise missiles,[3] and the "unexpected routes to potential disaster" arising out of complex technological systems such as the BrahMos.[4]

Also discussed were issues such as the absence of a broader set of standard operating procedures between India and Pakistan to deal with accidents involving sophisticated weapons systems,[5] the potential of the relative calm in the accident's aftermath as an opening for improving bilateral relations,[6] "the extent to which its (India's) offensive strike missiles are deployed in a launch-ready condition,"[7] and the role of luck in preventing escalation.[8]

Experts also weighed in on the commercial incentive for India to classify the accidental launch as human error rather than technical error,[9] Indo-Pakistani conflict (nuclear) dynamics,[10] whether the missile was launched by insiders with covert political authorization or due to an external cyber-attack,[11] and the policy options for the US amidst the changing strategic balance in South Asia[12] etc. The chapter undertakes a careful review of the documentary record from March to December 2022 to highlight the key issues discussed and the key actors and institutions involved in such discussions.

DOI: 10.4324/9781003543909-6

Serious Escalatory Risks of the Accident

Michael Kugelman, Director of the South Asia Institute at the Washington D.C.-based Wilson Center was quoted in a 11th March article by *Voice of America* warning of the serious escalatory potential of the accident, had it occurred during times of tension.

> the accidental strike shows how quickly an incident of this kind could escalate into something worse. This was a case of an Indian missile flying at supersonic speed across nearly 80 miles of Pakistani territory. Had this happened amid an actual security crisis, the escalation potential would be profound. And that's nothing to sneeze at, given that we're talking about two nuclear-armed states here," Kugelman told VOA. "Mistake or not, an incident like this has the potential to be escalatory in nature. The good news is this incident was handled well by both sides, to prevent any escalation.[13]

Kugelman elaborated further in an article for *foreignpolicy.com*, contending that the possibility of a deliberate Indian missile launch to test the interception capabilities of Pakistan's air defense system was remote.

> Had the Indian missile flown into Pakistan at a more fraught moment, the reaction from both capitals would not have been nearly as restrained... Islamabad likely lacks the capacity to shoot down such a fast-moving projectile, which Pakistani officials believe was a cruise missile. If something similar happens again, the same risks of escalation will come into play, given the possibility of casualties on the ground or in the air. Then there's the issue of the "technical malfunction" itself. India's missile fleet has had a strong safety record in recent years, which makes an egregious error surprising. This fact, coupled with India's lack of communication with Pakistan immediately after the launch, has prompted some Pakistani defense experts to raise the possibility that the launch was intentional—an operation to test Islamabad's response, including its ability to bring down such a missile....Given the high escalatory risks of such an operation, India would have little incentive to carry out such a reckless experiment.[14]

In an interview to *Rediff.com* on 15 March, Yogesh Joshi, a former MacArthur and Stanton Nuclear Postdoctoral Fellow at Stanford University and current research fellow at the Institute of South Asian Studies, National University of Singapore, averred that the missile misfire was a "big embarrassment" for a nuclear weapons state such as India. He speculated on the likely causes of the main reason cited by the Indian Ministry of Defence (technical malfunction) and the reason speculated on by strategic commentators (accidental launch). He also pointed to the 24-hour delay in the

DG-ISPR's statement as indicative of India having informed Pakistan about the missile accident.

> Generally, the principle followed is that an authorised launch always occurs and an unauthorised launch never occurs. This is what we call the always-never principle. Therefore, missile launch involves multiple fail safes – both mechanical and electronic – as well as rigorous procedures of command and control. Human resource training is equally intense. The accidental launch indicates that there has been some breach in any of these components....A technical error means that the entire stock of BrahMos missiles will have to be taken back to the assembly line and fixed for the said technical fault. A human error can be more idiosyncratic and one-off event....The fact that DG ISPR (Director General of Inter-Services Public Relations, the media and PR arm of the Pakistan Armed Forces) only issued its first statement 24 hours after the incident tells us that there was already a lot of discussion which had happened between the two sides before the world even knew of the incident....I think Pakistan was very mature in the way it dealt with the incident. India too owned up to the accident.[15]

Christopher Clary, Associate Professor of Political Science at the State University of New York and a non-resident fellow at the Stimson Center's South Asia Program, was also quoted by *Bloomberg* warning of the escalatory risks posed by high speed cruise missiles:

> Any cruise missile flight, especially for a faster than average cruise missile, such as BrahMos, occurs very quickly....even if Pakistani air defense radars observed the missile, which is by no means clear, there would have been only a few minutes or even a few seconds to coordinate a response to the intrusion.[16]

In a 17 March article for *War on the Rocks,* Clary expressed relief over the fortunate circumstances surrounding the accident, occurring as it did during a time of relative stability in India-Pakistan relations. He also highlighted the seriousness of the incident by pointing out that the accidental launch was the first time that a cruise missile had been launched from the territory of one nuclear weapons state into another. He criticized the Indian delay in communicating about the accident to Pakistan and the differing explanations advanced by the Indian government. Clary also made recommendations regarding prompt activation of secure lines of communication, addition of cruise missile flight test notifications to the existing 2005 ballistic missile flight test notification regime and re-programming of missile guidance systems with coordinates for unoccupied areas such as the Thar Desert instead of enemy targets.

> several media accounts indicated the dedicated army-to-army hotline was not used to inform Pakistan of the errant missile....Luckily, the apparently accidental launch occurred during a boring Wednesday evening for the subcontinent's sometimes fraught interstate politics. The 2019 Balakot crisis, though, offers a template of circumstances in which an errant cruise-missile launch could have proved catastrophic rather than merely embarrassing....While deadly military accidents were disturbingly common during the Cold War, last week's episode may be the first inadvertent launch of a cruise or ballistic missile by one nuclear power unto the territory of another nuclear power. The other disturbing characteristic of this episode is India's apparent lack of haste in communicating with Pakistan about the accident....Perhaps India hoped Pakistan would simply not notice, or that it wouldn't find the debris. Alternatively, perhaps India was uncertain as to the missile's trajectory and assumed that it had not strayed into Pakistan....adding cruise-missile flight test notifications to the existing ballistic-missile flight test notification regime between the two countries seems like a good idea....While the short range of some Indian and Pakistani nuclear-capable missiles likely precludes open ocean as a default target, both countries could declare that they will set the guidance systems of their weapons by default to unoccupied areas, such as the vast Thar desert.[17]

Cruise Missiles to Be Brought under 2005 Agreement, Deliberate Launch to Test Pakistan

Sumit Ganguly, the Rabindranath Tagore Professor of Political Science at Indiana University,[18] commended the "restraint" of the Pakistani military in the *South China Morning Post*. Taking note of the scant information regarding the institutions involved in India's nuclear force structure and recent evidence suggesting the mating of some warheads with delivery vehicles, Ganguly recommended that the scope of the ballistic missiles pre-notification agreement of 2005 be expanded to include cruise missiles.

> There is little information in the public domain about the structure of India's nuclear command authority because of an acute penchant for secrecy about security matters. Based on public statements from prominent former officials and a handful of news releases, it appears that the nuclear force structure lies in the hands of civilian authorities. Furthermore, these statements also suggest that extensive drills are routinely conducted to ensure that the nuclear apparatus can respond with alacrity to any imminent threat. Initially, the atomic energy establishment had day-to-day custody of the fissile core of the nuclear warheads while warheads and missiles were the preserve of the uniformed military. More recently, however, evidence has emerged that missiles have been

> mated with warheads in a segment of the nuclear force....Merely stating that a "technical malfunction" led to the launch will do little to reassure a domestic audience let alone an adversary that is prone to assume the worst of its principal enemy....Since an accord on the pre-notification of ballistic missile launches is already in place, it may be in the interest of both parties to expand its scope. This would require either party to promptly and forthrightly inform the other in the event of an accidental launch. Since a nuclear hotline has existed since 2004 between the two highest foreign policy bureaucrats, its ambit could be broadened to include the issue of accidental missile launches.[19]

Hussain Haqqani, former ambassador of Pakistan to the US (2008–2011), Director for South and Central Asia at the Hudson Institute and a Pakistani citizen based in the US,[20] warned that the accidental missile launch was further evidence of the acute need for regular communication between nuclear-armed India and Pakistan in an 18th March article published in *The Hill*.

> India and Pakistan both possess nuclear weapons, are constantly suspicious of each other and have fought four wars since their Independence... In this instance, Pakistan's restraint and India's belated admission of its error avoided the worst fallout. But India's decision to set up a high-level Court of Enquiry should not suffice for the matter to be set aside. Pakistan has a right to know what happened and what mechanisms India is now putting in place to avoid the recurrence of similar accidental missile launches....Nuclear-armed rivals need to communicate, if for no other reason than to avoid catastrophic consequences of accidents and misunderstandings. Its reservations about Pakistan's support for jihad in Afghanistan and Jammu and Kashmir notwithstanding, India still needs to engage with Pakistan over matters such as its 9 March accidental missile launch.[21]

Hassan Aslam Shad, an international lawyer and a Harvard Law School graduate, channeled the concerns of his sources in the Pakistan military in an article for *The National Interest* in which he worried as to whether India had deliberately fired a cruise missile into Pakistan.

> As global attention remains fixed on the ongoing Russia-Ukraine war, South Asia just averted a catastrophic war between two nuclear-armed neighbors....And, to its credit, despite tracking the object since its launch from India and until it detonated in Pakistani territory, Pakistan did not exercise its right to retaliate in self-defense....To date, India has not confirmed whether the missile was the nuclear-capable BrahMos or explained how a "technical malfunction" could have caused an accidental launch.

> Pakistan has asked for a joint probe into the incident....sources in the Pakistan military have informed me that Pakistan has not ruled out the possibility that the launch was an intentional act to gauge Pakistan's war readiness and/or ability to track India's BrahMos. If anything, Pakistan's tracking the missile from its launch until its detonation has demonstrated Pakistan's capability to track and counter such attacks.[22]

Fahd Humayun, a PhD candidate at Yale University's Department of Political Science, examined the many unanswered questions surrounding the missile accident, including that of a deliberate launch in a 25 March article for the Jinnah Institute. Viewing the restrained response by Pakistan to the missile accident as a continuation of its similarly restrained reaction to the Indian airstrike at Balakot, Humayun called for the reaffirmation of crisis communication procedures between the two countries.

> The missile landed 124 kilometers deep into Pakistani territory and it is a miracle that it did not escalate into a full-scale crisis between two highly crisis-prone nuclear armed adversaries. Troubling questions abound. Why did the system's failsafe mechanisms not work? If the malfunctioning wasn't mechanical, what explains the ostensible lapse in operational, command and control procedures? Why didn't Indian authorities track the missile after takeoff, and if they did, why did they not immediately establish contact with Pakistan? And most worryingly, could this in fact have been a deliberate attempt to measure Pakistan's responsiveness and air defenses, in other words a disguised flex, as India doggedly expands its military capabilities? The answers are unclear. What is clear is that India's capacity to safely handle sensitive technology in South Asia's nuclear environment is now rightly being called into question...While Pakistan's response to this averted crisis echoes the maturity of sensibly demonstrated restraint to the Balakot provocation in 2019, the need for a comprehensive and transparent investigation into the misadventure of March 9th under international oversight, along with bilaterally reaffirming crisis communication protocols, is paramount.[23]

Lack of Standard Operating Procedures, Pakistan's "Dual-Use" Threat Perception

Debak Das, Assistant Professor in Peace and Security at the Joseph School of International Studies at the University of Denver and an affiliate at the Freeman Spogli Institute for International Studies at Stanford University,[24] worried in a 25 March article in *The Bulletin of the Atomic Scientists* that lack of standard operating procedures for crisis communication risked further lowering the already low threshold for India-Pakistan military engagement. Das also pointed out that although there was no conclusive evidence that the BrahMos was a part of India's nuclear forces, reports of it claimed nuclear-capability in the Indian media and the resulting possibility of India nuclear-arming the missile as part of a counterforce nuclear doctrine would affect Pakistani threat perceptions.

> The missile at the center of the accident was reportedly a surface-launched supersonic BrahMos cruise missile jointly developed by India and Russia with a range of 290 kilometers. India is also developing an extended air-launched variant with a range of 800 kilometers. The BrahMos is not a part of India's nuclear forces, though it has been reported to be nuclear-capable. India may realize its nuclear potential if it moves toward a counterforce nuclear doctrine, according to some scholars....In recent years, the threshold for military engagement between India and Pakistan has been reduced....If the missile had been armed, Pakistan might have responded with military force. Pakistan reportedly prepared to launch a proportional retaliatory strike against India before it realized that that the missile might have been accidentally fired....The agreement (on the pre-notification of flight testing of ballistic missiles) concerns only ballistic missiles and not cruise missiles like the one involved in the March 9 accident. Further, even if the agreement concerned cruise missiles, an accidental launch like the one on March 9 does not provide time for advance warning....neither country reportedly used the (DGMO) hotline during the recent accident. There is a stark lack of standard operating procedures between the two states to deal with accidents related to sophisticated conventional weapons (including both ballistic and cruise missiles) that could lead to military escalation.[25]

The Stimson Center, a Washington-based think-tank, published a three-part series on the accidental missile launch with contributions from Indian, Pakistani and American authors featured in its "South Asian Voices" (SAV) section.[26] Frank O'Donnell, Deputy Director and Fellow at the Center's South Asia Program, reckoned in a 25 March interview with SAV Managing Editor Brigitta Schuchert and Associate Editor Isha Gupta, that the problem with the missile accident seemed to lie in the human or technological element within the Indian command and control system responsible for authorizing the missile's initial launch. O'Donnell also dwelled on the increased risk stemming from Pakistan's perception of the BrahMos as a dual-use weapon system with some missiles being part of India's nuclear forces.

> The more common type of missile test failure, in most states, is when the missile is launched according to plan, but when a component malfunctions mid-flight, and the missile veers off course and is then destroyed

> in-flight by the test managers. This test failure can be due to malfunctions in various parts of the missile, for example its navigational or propulsion systems. However, the phrasing of the Press Information Bureau announcement, of "accidental firing," suggests a more concerning type of missile failure, relating to some human or technological element of the command-and-control system authorizing the initial launch of the missile. Even if the Prime Minister's Office and Ministry of Defense had given their prior approval for the test-launch, and the Indian Air Force had scheduled the test for that day, the nature of this malfunction still qualifies as an unauthorized missile launch. Symbolically, this is a technologically leading missile in India's arsenal, and is its only supersonic cruise missile....It is strategically important as Pakistan believes that BrahMos missiles are dual-use, and that some missiles may be part of India's nuclear forces. Even if not, there have (been) strong indications that Indian military planners see the BrahMos as effective for conventional counterforce missions. There will be an option for the test managers to order the missile to self-destruct as soon as the unauthorized launch was detected, just as there would be for a launch which veered off course...There was more than enough time for the Indian test managers to react and order the missile to self-destruct within the first minute after launch.[27]

Ali Ahmed, a freelance blogger[28] and author of books such as *India's Doctrine Puzzle: Limiting War in South Asia* and *On War in South Asia*, analysed in a 25 March article in the SAV section that the reason for Pakistan's decision to not retaliate to the errant missile was to improve its relationship with India. Highlighting the similarity of the positions of both countries on the Ukraine war and the mature handling of the missile crisis by both countries, Ahmed averred that it would enable them to make progress on other contentious issues holding back normal relations.

> Though the supersonic missile did little damage, the knowledge that it was a "dual-use" BrahMos missile could have proven a trigger for crisis escalation. Their similar positioning on Ukraine comes at the heels of growing cooperation in other arenas, which, when coupled with their non-confrontational attitude to the accidental missile launch, can potentially lead to this positive turn in their relationship becoming a trend. Pakistan also avoided retaliating against the accidental missile launch since it is interested in improving its relationship with India now that its reaction to India's revocation of Article 370 has run its course....The maturity of India and Pakistan in the wake of the misfired missile signals improved relations and reflects their growing impetus to strengthen CBMs (Confidence Building Measures), enabling both to move in tandem on other contentious bilateral issues.[29]

Amina Afzhal, a Canada-based researcher with more than ten years of work experience in Pakistani think-tanks,[30] reiterated the concerns of Pakistani security experts regarding Pakistan's air defense capabilities to intercept a future intentional Indian missile strike. She stressed that the lack of escalation in the aftermath of the missile crisis was primarily due to good fortune[31] and that the failure to use the military-to-military hotline could have more serious consequences in a future conflict.

> Several media reports claim that India never used the military-to-military hotline following the "inadvertent" missile launch, which would have provided a quick communication link between the two countries designed to reduce miscalculation and aimed at risk mitigation. The failure to use the hotline could imply more serious consequences for both India and Pakistan during a future conflict. Other reports have suggested the possibility of a Pakistani retaliatory strike. Pakistan's subdued reaction has been attributed to the realization that "something was amiss."...While some in India commended the Pakistani military's "muted" response that potentially helped avert a crisis, others have criticized Pakistan's failure to intercept the missile's launch. Observers in Pakistan have also raised questions about the adequacy of Pakistan's response and the country's ability to defend itself against an Indian attack. Like India, Pakistan will also have to address issues related to public confidence amid the possibility of radar coverage and air defense gaps. Since good fortune seems to have played a greater role in ensuring a low-key Pakistani response following the incident, both Pakistan and India need to revisit their bilateral relations, especially in the nuclear realm.[32]

Writing in *Arms Control Today*, Daryl G. Kimball, the Executive Director of the Arms Control Association,[33] appeared to misconstrue the Agreement between India and Pakistan on Pre-Notification of Flight Testing of Ballistic Missiles of 2005 as also applying to cruise missiles such as the BrahMos.

> An unnamed Pakistani official told *Reuters* on March 12 that the missile was a nuclear-capable, land-attack BrahMos cruise missile jointly developed by Russia and India. The BrahMos has a range of 300 to 500 kilometers. The incident raises questions not only about India's operational safety procedures and controls, but also the extent to which its offensive strike missiles are deployed in a launch-ready condition....The Indian government took 48 hours to officially confirm the misfire despite a 2005 agreement that requires each country's defense ministry to give its counterpart at least 72 hours' notice before conducting a ballistic missile flight test.[34]

A Deliberate Launch to Test Pakistan, Launch Authority Delegated to Lower Chain of Command

In an article for the *Atlantic Council* on 5 April, Riaz Khokhar, a research associate at the Center for International Strategic Studies (CSIS) and a former Asia Studies Visiting Fellow at the East-West Center in Washington, and Asma Khalid, a former South Asian Voices Visiting Fellow at the Stimson Center, speculated that the missile accident should be understood as an unauthorized launch and not as a technical malfunction as it seemed to be fired by people in the lower chain of command with the ability to remove the mechanical and electronic locks and with access to the launch codes.

> based on publicly available data on the subject and our own primary interviews, we argue that there is a far greater likelihood that the missile was an unauthorized launch on part of the missile crew of the Indian Air Force (IAF). Recent news reports in India have found that a group captain of the IAF and his colleagues were allegedly involved in the missile launch....The group captain had charge of a Mobile Command Post, which controls four Mobile Autonomous Launchers (MALs)—a complex, land-based BrahMos missile system. Depending on the input of the commander-in-charge, the MAL can fire the missiles towards designated targets within four minutes. To clarify, the BrahMos flies on a "Fire-and-Forget" principle and cannot self-destruct once launched.....nuclear scientists consider only those launches as accidental that occur because of a technical error. If it is fired by people in the chain of command with access to the codes but lacking political authorization, it would constitute an unauthorized launch. Moreover, in the past, when India kept its warheads and their delivery vehicles separate (known as "de-mated"), it incorporated "super permissive action links (PALs)," which are mechanical and electronic codes to avoid unauthorized and accidental launches. With increasingly higher readiness levels of India's conventional and strategic forces, one can expect far more stringent safety and security protocols.....With far greater safety protocols than previously installed "super PALs," one can only assume that the commander in charge must have removed the mechanical and electronic locks and used the launch codes to fire the missile. There appears to be no other possible explanation unless India's sophisticated missile system lacks rigorous safety and security protocols. In other words, India's lower chain of command apparently knows the launch codes, and it would be extremely dangerous in the case of missiles mated with warheads and other canisterized (permanently armed) missiles. Additionally, it reflects poorly on the training and psychological assessment of the Indian armed forces personnel and raises questions about their intent.[35]

Khokhar and Khalid then asserted that the unauthorized launch by the IAF personnel was meant to test Pakistan's detection and interception capabilities, claiming that their views represented those of a large section of the Pakistani strategic community. They also channeled the views of a senior Pakistani government official regarding the reasons for Pakistani restraint in the aftermath of the Indian missile misfire.

> We argue that IAF personnel might have tried to ensnare Pakistan into mobilizing military forces or simply check the BrahMos missile's capability to dodge Islamabad's air defense system. Since IAF personnel arguably operated with a bad motive, they did not activate the military hotline of communication to inform Pakistan. In case of an escalation of the crisis, they likely expected the political leadership in New Delhi to manage and respond....In this context, a larger segment of Pakistan's strategic community whom we interviewed or whose comments we read, feels that the fired missile was not accidental and that it was instead a deliberate attempt to assess Islamabad's response spectrum, the types of conventional or strategic forces that Pakistan deploys, and the organizational level at which decisions are made. India could make effective use of this information to materialize its flirtation with counterforce strikes against Pakistan's strategic forces and command and control system in the future....the restraint was due to the fact that there was no ongoing crisis between the two countries, not necessarily because of consistent and effective surveillance of the missile and its ground-based stations. In a crisis, Pakistan is unlikely to show this restraint and risk a potential nuclear strike on its soil.[36]

Strafasia, a UK-based online publication for strategic affairs featured an article by Dr. Adil Sultan, Head of Department of Defense and Strategic Studies (DSS), Air University Islamabad, in which he analyzed that India's stated reason of a "supervisory lapse" was even more worrying than a technical malfunction as it pointed to a fundamental problem in the "command and control architecture." He also dismissed contentions that the motive for India's missile launch was a deliberate testing of Pakistan's capabilities and an attempt to map its air defense system. Sultan also drew a worrying conclusion from his analysis, that the launch authority for India's strategic missiles had been "pre-delegated" to the lower chain of command with no real-time oversight from the political leadership.

> On March 9, 2022, a super-sonic BrahMos cruise missile crashed inside Pakistani territory that India later claimed was an 'accidental firing' and a technical malfunction during routine maintenance....Subsequent reports appearing in the media, however, indicate that the incident may not be a

> result of a technical glitch but a supervisory lapse where a Group Captain rank officer in charge of the mobile command post was held responsible for the incident. If true, this is more disturbing than a technical malfunction as it exposes serious gaps in India's command and control architecture which may not be geared to preclude the possibility of an accidental or unauthorized launch....the likelihood of a 'bolt from the sky' kind of a pre-emptive strike would still need a brewing crisis and some justification, which was absent in the recent episode, allowing the Pakistani side to conclude that the missile was not part of a pre-emptive 'first strike' and could have been an unintended missile launch that went rogue. The air defense mapping theory also does not hold much credence as India has several other means available to locate Pakistan's radar deployments. India's network of satellites provides adequate situational awareness... It has a bilateral agreement with the US known as Basic Exchange and Cooperation Agreement (BECA), which provides India access to classified real-time signal intelligence (SIGINT) and other sensitive information gathered by the US satellites about India's Northern and Western borders....However, looking at the flight path of the missile it could be concluded that India's strategic missiles are operationally deployed in a higher readiness status with pre-fed coordinates against targets inside Pakistan, and the launch authority has already been pre-delegated to a relatively lower level rank with no central oversight that could have prevented such incidents.

Khan also outlined the reasons for India's defective command and control structure, which he attributed to the discordance between its politicians, military and scientists.

> The lack of oversight seems to be a result of prevalent dissonance amongst India's strategic enclave (political, military and scientists). The political leadership considers it to be their prerogative to maintain strict oversight over India's strategic and nuclear capabilities with little or no understanding of how these could be employed to achieve the desired political objectives. The disconnect in India's NFU commitment amongst senior political leadership is one such example. The military leadership remains incoherent due to inter-service rivalries and also because they are generally kept outside the nuclear and strategic decision-making process by the political leadership, primarily to limit their influence. The scientific community, on the other hand, continues to build solutions for unspecified threats, including the development of weapons ranging from short range to inter-continental ballistic missiles, without taking into consideration the military needs. This visible dissonance amongst all the three components (political, military

> and scientists) makes it difficult to ensure a coherent and fail-safe C-2 structure.[37]

Accidents Inevitable in Complex Systems, Design Shortcomings, Surprise-Onset Crises

Dr. Zian Mian, physicist and co-director of Princeton University's Program on Science and Global Security,[38] and Dr. M. V. Ramana, a Professor and Simons Chair in Disarmament, Global and Human Security at the School of Public Policy and Global Affairs (SPPGA), University of British Columbia, situated the missile accident in a *Scientific American* article in the context of a series of previous accidents involving military systems. They also cited influential scholarship to warn that such accidents were inevitable in complex weapon systems.

> The firing of the BrahMos missile falls within a long history of accidents involving military systems in India. Military aircraft have strayed across the borders during peacetime. India's first nuclear submarine was reportedly "crippled" by an accident in 2018, but the government refused to divulge any details. Secrecy has prevented the investigation of an apparent failure of India's ballistic missile defence system in 2016. Engagements between India and Pakistan can arise from such accidents, as in 1999 when a Pakistani military plane was shot down along the border by India, killing 16 people. Pakistan has had its share of accidents, including a Pakistani fighter jet crashing into the capital city in 2020. All these weapons systems are inherently accident-prone because of two characteristics identified by organizational sociologist Charles Perrow decades ago—interactive complexity and tight coupling—that combine to make accidents a "normal" feature of the operation of some hazardous technologies. The first characteristic refers to the possibility that different parts of the system can affect each other in unexpected ways, thus producing unanticipated outcomes. The second makes it hard to stop the resulting sequence of events. For Perrow, "the dangerous accidents lie in the system, not in the components," and are inevitable.[39]

Fang Xiaozhi, Associate professor at the School of International Relations, PLA National University of Defense Technology, surmised in a 20 April article in *China Military Online*, the only official English-language news website of the People's Liberation Army (PLA)[40] that the missile accident exposed serious shortcomings in design and in R&D.

> India's current status of weapons research and development still lags behind its ambitious development plan, especially in the field of missile research and development. Although India has mastered relevant

> technologies of missile manufacturing, it has few independent intellectual property rights, and many core technologies come from abroad. In addition, India has failed to establish a complete defense industrial system, and the weapons it has developed are always unsatisfactory. The "misfiring" of the BrahMos supersonic anti-ship missile this time also fully exposed the shortcomings of its missile design and the lack of missile research and development capabilities. India's missile technology is still facing multiple difficulties, and it still has a long way to actual combat.[41]

A 17 May report by the Senior Study Group of the United States Institute for Peace (USIP) titled "Enhancing Strategic Stability in Southern Asia" contextualized the missile accident as a worrisome development with serious escalatory potential and as the trigger for a "surprise-onset" crisis.

> Even though Southern Asian nations are confident in their ability to manage escalation, this is the only place on earth where a nuclear power has conducted an airstrike on the territory of another nuclear power, as India did after the 2019 Pulwama terrorist attack. It is the only place where soldiers of two nuclear powers have attacked and killed one another at their disputed border, as Chinese and Indian soldiers did in 2020 in the Galwan Valley. It is also the only place where one nuclear power has accidentally launched a missile into the territory of another, as India did in 2022....The missile was unarmed, caused no casualties, and came during a period of comparative diplomatic calm between the two sides, but still underscores the potential for surprise-onset crises in the region.[42]

Sameer Lalwani, a senior fellow for Asia strategy at the Stimson Center and a nonresident fellow with the Sigur Center for Asian Studies at George Washington University, reckoned in a 17 May article that US assistance and "more reliable de-escalation measures" were required for crisis-stability in South Asia.

> Luckily, this incident occurred in peacetime, amid a cease-fire agreement between the two rivals that has proved surprisingly durable for 15 months. Reports later surfaced that Pakistan had prepared a "retaliatory strike," but held back. If such an accidental missile launch occurred during the most recent India-Pakistan crisis in 2019, it could have been catastrophic. Mishaps that occur during the fog and friction of a militarized crisis can easily escalate. And accidents endemic to large, complicated military organizations—like a rogue missile launch, a false missile attack warning or the shoot down of a commercial plane—can rapidly spiral out of control. The incident exposed the risk for dangerous accidents with advanced weapons systems and a concerning inability or unwillingness

> to quickly alert a rival to the accident, even through an existing hotline. Rather than waiting until the next crisis to spring into action, the United States can help both sides develop their own crisis management tools now...including enhanced intelligence, communication links, information sharing and dialogue venues.[43]

US Strategy to Maintain Stability in South Asia, Deliberate Launch and Hindutva Ideology

Daniel Markey, Andrew Scobell and Vikram Singh, all co-Chairs of the aforementioned USIP Senior Study Group on Strategic Stability in Southern Asia, analyzed that the US should help maintain peace in South Asia by undertaking efforts in three key domains: 1) management of core regional disputes, 2) facilitating strategic regional stability and 3) effective management of potential crisis involving nuclear-armed countries in the region.

> The United States should also encourage India and Pakistan to consider unilateral or bilateral steps, such as renouncing specific technologies like nuclear depth charges and adding cruise missiles to the 2005 missile test pre-notification agreement. Such moves would both help reduce the use of especially destabilizing technologies and build confidence for more significant arms control discussions.[44]

In a podcast for the USIP on 19 May, Markey erroneously explained the missile accident as an "accidental launch by India of a hypersonic missile which is considered dual-capable."[45]

Anum Khan, Associate Director at the Center for International Strategic Studies (CISS), Islamabad and a former international relations analyst for Pakistan's Strategic Plans Division, contended in a 28 June article for the Seoul-based, The Asia-Pacific Leadership Network for Nuclear Non-Proliferation and Disarmament (APLN) that the accident was indicative of Indian missiles already being deployed in "launch on command" positions facing Pakistan. After explaining the Indian nuclear chain of command and the multiple steps required for authentication of an order to launch, she also highlighted the possibility of an insider attack or a cyber attack as causes for the missile accident.

> According to the 2017 Joint Doctrine for the Armed Forces, the authorisation of the launch of nuclear weapons can only be done by the Prime Minister who heads India's Nuclear Command Authority (NCA). The launch order goes directly to the Indian Strategic Forces Command. Nevertheless, Vipin Narang states that, "the Prime Minister may not be able to release nuclear weapons without the National Security Adviser, who is the interface between the NCA and the Strategic Forces Command."

> Critics within the Indian military have expressed doubt regarding India's operational capacity saying the military is not adequately involved in nuclear decision-making. The BrahMos missile that accidentally landed in Pakistan was without a warhead, but what is notable is that the concerned missile is dual-capable, that can also carry a nuclear warhead....An unauthorised launch of nuclear capable missiles is, however, not as simple as pressing a button. Agreed codes need to be sent over several separate communication channels to ensure that the officer in-charge can confirm if it is an authentic order...The BrahMos missile crisis of March 2022 also suggests that nuclear-capable Indian missiles are already deployed in launch on command positions vis-à-vis Pakistan....Hence, there needs to be a thorough investigation about the recent missile launch, to ascertain whether it was perpetrated by insiders through orders by those in the chain of command with access to the launch codes, or by outsiders, via a cyberattack for instance.[46]

Aarish U. Khan, a research analyst at the Institute of Regional Studies, Islamabad surmised in 29 September article in *The Diplomat*, a US-based magazine, that the Pakistani policy elite's suspicion of the missile firing being a deliberate event could not be divorced from its "anxieties about India's surging Hindutva ideology.....permeating into the minds of policymakers and policy influencers in Pakistan."

> Speaking at an event jointly organized by the Islamabad-based Centre for International Strategic Studies and the London-based International Institute for Strategic Studies in Islamabad, former Director General of the Strategic Plans Division of Pakistan Lt. Gen. (Retd.) Khalid Ahmed Kidwai said that the launch of an Indian missile into Pakistan earlier this year was "no accident."...Kidwai argued that missiles could not be accidentally launched into a neighboring country as if they were bullets fired from infantry rifles. He added that the coordinates of missile trajectories were pre-fed into the system on the basis of high-level approvals. This, he asserted, required extensive preparation that could not occur on the spur of the moment, implying that the Hindutva-inspired Indian government had given the required high-level approvals before the missile was "deliberately" fired into Pakistani territory. The problem with this line of argument, however, is that the Indian court of inquiry did not deny the prior feeding of the waypoints into the missile's path of trajectory – which, according to the Indian authorities, is done by countries as a war-preparedness measure, to the extent that missiles need only minor last minute adjustments to hit the exact targets....The pertinent question, however, is not whether Kidwai's assertion passes the test of technical analysis, but where such anxiety is originating from. He is not alone in believing this

missile launch was a deliberate act by an Indian government inspired by an aggressive nationalistic ideology. This perception might have to do not only with the recent Hindutva-inspired developments inside India, but also the provocative pronouncements of senior Indian ministers with regard to Pakistan.[47]

Entanglement, Strategic Risk Reduction Centers, India's Integrated Rocket Force

In a 31 March online panel discussion titled "Skirting the Nuclear Threshold: Conflict De-escalation and Strategic Stability in South Asia," organized by the New Delhi-based Council for Strategic and Defence Research, Feroz Hassan Khan, a Research Professor in the Department of National Security Affairs of the US Naval Postgraduate School, a former Brigadier in the Pakistan Army and Director of Arms Control and Disarmament Affairs in Pakistan's Strategic Plans Division, weighed in on the implications of the missile accident. Khan began by averring that the missile intrusion was viewed as a sole freak accident by sections of the Pakistani strategic community. He explained that the accident had raised doubts over the hitherto presumed improbability of a 'bolt out of the blue' strike in South Asia although it also provided a window into how arsenals are maintained in peacetime. Khan further pointed out that other sections in the Pakistani strategic community believed that the accidental launch was a deliberate Indian move to test Pakistani defenses or to find weaknesses. He attributed Pakistan's restrained response to its preoccupation with domestic politics and to its disappointment over the international community's dismissive response.

Khan also analyzed that the accident raised questions over what would happen in case the accidentally launched missile had real warheads, especially canisterized warheads. He informed that multiple questions were being asked in Pakistan as to whether India was immediately aware that the missile went astray and whether there was a self-destruct mechanism on the missile.

Reminding his fellow panelists that this was the first time a missile had been accidentally fired from the territory of one nuclear armed state into another, Khan criticized this as unacceptable. Commenting on the limitations of hotlines, he characterized them as necessary but not sufficient and gave the example of the Kathmandu hijack incident in which the Pakistani DGMO (in Khan's presence) tried in vain to inform the Indian DGMO using the hotline but couldn't get through in time as the latter was supposedly playing golf.

Identifying several developments with "tremendous potential for inadvertence" such as the accuracy of new missile systems, increase in cyber capabilities, unmanned UAV's, sea-based deterrence and NC3 systems,

Khan called for the establishment of Strategic Risk Reduction Centres as "central nodes" that could be in constant contact with the other side for immediate notification of accidents such as the missile misfire.

He also agreed with fellow panelists that the nuclear-conventional entanglement issue (in cases such as the BrahMos with its dual-use capability) was a serious one that needed to be discussed at the professional (military) level to generate confidence. Khan also highlighted the lack of interception options for countries facing an incoming, supersonic missile with aircraft not being capable of being scrambled and ballistic missile systems only protecting major economic and strategic command centers but not being economical to protect the entire territory.

Khan concluded by calling on India and Pakistan (as also India and China) to formalize a de-alert/non-alert agreement to reduce the risk of a sudden-onset security crisis and unwanted conflict escalation. He also recommended a political-military-strategic dialogue between the two countries at two separate levels: military and political. Lamenting the absence of recent discussions on doctrinal clarity, Khan recounted how India and Pakistan discussed security concepts a lot more before 1999 in the wake of the 1998 tests by both countries.[48] Khan also articulated his views in a paper in the *Journal of Peace and Nuclear Disarmament* in which he starkly warned that the risk of a South Asian security crisis was increasing because of the "entanglement of dual-capable missile systems and assortment of disruptive technologies, now in the inventories in both dyads (India-Pakistan and India-China)."[49]

More recently, Daniel Markey, the senior adviser on South Asia at the United States Institute of Peace, expressed concern that the risks to South Asian stability had increased due to geopolitical changes, growth in nuclear arsenals and advancements in missile technologies. He warned that accidents involving human and/or technical error were inherent in the operation of such complex technologies and that the potential for escalation was heightened by the lack of timely use of reliable communication links between India and Pakistan.

> As a USIP senior study group report concluded last year, the problem of nuclear stability in Southern Asia is getting harder to manage because of geopolitical changes, such as rising India-China border tensions, as well as evolving military technologies, including growing nuclear arsenals and more capable delivery systems...Accidents do happen. India's misfire of a BrahMos missile test into Pakistan last year proved that point perfectly. No matter how well designed, nuclear systems are complicated and involve the potential for human or technical error. When something does go wrong, overreaction by opposing forces is less likely when they have a greater degree of confidence in, and knowledge of, the other side. Reliable

> and secure communications—in the form of hotlines—can help, but only to the point that they are actually used in a timely manner. Apparently, India failed to do so during the BrahMos incident. A return to serious India-Pakistan crisis could be just one terrorist attack away....When India arms itself to deter China, Pakistan perceives new threats from India and will likely pursue enhanced capabilities of its own. In a worst-case scenario, Southern Asia could be entering an accelerated nuclear arms race in which uneven waves of new investments in capabilities and delivery systems will alter perceptions of deterrence and stability in dangerously unpredictable ways.[50]

Debak Das, Assistant Professor in Peace and Security at the Joseph School of International Studies at the University of Denver, recently brought up the example of the missile accident to warn about the dangers of inadvertent escalation resulting from India's decision to create an Integrated Rocket Force (IRF) which is likely to include both nuclear-capable and conventional missiles within the same force.

> New Delhi is creating an Integrated Rocket Force (IRF) that will contain both nuclear-capable and conventional ballistic and cruise missiles... India's IRF is being developed largely in response to threats posed by China and is expected to give India a larger array of options to conduct "non-contact warfare" through the use of conventional ballistic and cruise missiles in stand-off strikes.India's creation of the IRF is a mistake. The IRF will likely place nuclear-capable and conventional ballistic missiles within the same force; this will change India's present policy of ballistic missiles being dedicated to only carrying nuclear weapons. Using the same missiles in both conventional and nuclear roles increases the chance of inadvertent escalation. An adversary will not be able to tell if an incoming missile is armed with a nuclear weapon or a conventional payload. The inability to distinguish between an incoming conventional and nuclear strike may lead to preemptive nuclear retaliation from the other side...The risk of an inadvertent missile launch is not hypothetical. In March 2022, India accidentally fired a BrahMos cruise missile into Pakistan. The missile was not armed, and Pakistan did not respond. However, given the flight time of about six minutes—in which Pakistan had to decide whether the incoming missile was armed or not—there was a high risk of a Pakistani retaliation, even before the incoming BrahMos struck land. Importantly, India has a declared conventional role for the BrahMos missile (which it co-developed with Russia), which might have had a calming influence on the Pakistani response. In a wartime environment, the response might have been different. The lesson from this accidental launch is that nuclear and conventional missiles should be kept separate to prevent inadvertent escalation.[51]

Manuel Herrera, a researcher in the Multilateralism and Global Governance programme at the *Istituto Affari Internazionali* (IAI), an Italian international relations think-tank, also called for strengthening "information sharing and establishing timely reporting mechanisms to avoid the recurrence of such incidents and to prevent miscalculation."[52] He also recommended the resumption of dialogue on respective missile forces within the framework of the 2005 pre-notification agreement.[53]

Analysis

In the aftermath of the missile accident, several international experts mostly based in the US (with only one from China) attempted to understand the intricacies of the accident, its regional consequences and ramifications for US foreign policy. The discussions among international experts were similar to the discourse of the Indian and Pakistani strategic community while also having some key differences.

The similarities included a quick realization of the implications of the missile accident for South Asian security, despite the missile being unarmed and causing no casualties in Pakistan. Of particular salience was the clearly articulated understanding that this was the first case of a nuclear armed state firing a cruise missile into the territory of another nuclear armed state with which it shares a disputed border. The occurrence of the missile misfire during a period of relative calm in the India-Pakistan relationship did little to assuage the concerns of international experts regarding a similar accident with a missile happening during a period of high tension (such as in the aftermath of the Balakot airstrike by India), resulting in a Pakistani retaliation and triggering a dangerous escalation.

The aforementioned nightmare scenario provided the context in which the Indian delay in acknowledging the accident and its eschewal of the military hotline to inform Pakistan also came in for criticism with questions raised about the adequacy of standard operating procedures and about the reliability of the command and control system. There was also speculation regarding (from experts of Pakistani origin) whether the missile launch was a deliberate attempt to test Pakistan's monitoring, interception and response capabilities. Worries were also expressed over Pakistan's perception of the BrahMos as nuclear capable (and the possibility of India nuclear-arming the missile as part of a counterforce nuclear doctrine and as part of an Integrated Rocket Force) lowering the threshold for military conflict between the two nations in case of future accident.

Some experts also contended that the accident was better understood as an unauthorized launch with the worrying implication that India's lower chain of command had access to the launch codes and had the capability to remove the various mechanical and electronic locks. One Chinese expert cast aspersions on India's missile research and development capabilities and the resulting design defects that led to the accident.

International experts made a range of recommendations to ensure prompt communication and escalation prevention in case of such accidents. These included inclusion of cruise missiles under the ambit of the 2005 agreement on pre-notification of ballistic missiles, eschewing development of entangling institutions such as the Integrated Rocket Force, creation of Strategic Risk Reduction Centers, resumption of dialogue at the political and military levels etc. Cognizance was also taken of remarks by Lt Gen (Retd) Khalid Kidwai, former Director General of Pakistan's Strategic Plans Division (SPD) and Advisor to its National Command Authority, contending that India's launch was the deliberate handiwork of its Hindutva ideology inspired government.

The international strategic discourse on the accident was different from the domestic strategic discourses in India and Pakistan in the manner in which the causes of the accident were understood by a section of international strategic experts. It was reckoned that accidents such as the March 2022 one were inevitable in complex systems due to numerous sub-systems and the unforeseeable number of potential interactions between them at one level and between these sub-systems and fallible humans on another. The implication of this line of analysis was that incremental changes to standard operating procedures and affixing responsibility on culpable missile unit operators (through the Indian Court of Inquiry) would not solve the problem of irreducible complexity that would continue to pose unavoidable accident risks to missile arsenals worldwide, including those in India and Pakistan.

Another key differentiator of the international strategic discourse was the prescriptions made by American strategists for US foreign policy in terms of recognition of the recurring potential for such "sudden-onset crises" in a regional milieu with two nuclear-armed neighbors with tense relations, large military organizations, intense distrust and minimal communication. The prescriptions spanned the gamut of proactive crisis management strategies from enhanced intelligence to institutionalization of more dialogue forums.

Other unique observations included by international experts included pointing out the limitations of currently existing hotlines in ensuring reliable, real-time communication amidst the crowded engagement calendars of the two DGMOs. The potential of new technological developments to cause inadvertent accidents was also highlighted including from enhanced cyber capabilities, new missile systems, Unmanned Aerial Vehicles (UAV's), NC3 systems and sea-based deterrence capabilities.

Notes

1 Frank O'Donnell, "SAV Q&A with Frank O'Donnell: Key Takeaway's from India's Missile Malfunction," *South Asian Voices,* March 25, 2022, accessed

September 3, 2023, https://southasianvoices.org/sav-qa-with-frank-odonnell-key-takeaways-from-indias-missile-malfunction/

2 Christopher Clary, "The Curious Case of the Accidental Indian Missile Launch," *War on the Rocks*, March 17, 2022, accessed October 10, 2023, https://warontherocks.com/2022/03/the-curious-case-of-the-accidental-indian-missile-launch/

3 Frank O'Donnell, "SAV Q&A with Frank O'Donnell: Key Takeaway's from India's Missile Malfunction," *South Asian Voices*, March 25, 2022, accessed September 3, 2023, https://southasianvoices.org/sav-qa-with-frank-odonnell-key-takeaways-from-indias-missile-malfunction/

4 M. V. Ramana and Zia Mian, "India's Inadvertent Missile Launch Underscores the Risk of Accidental Nuclear Warfare," *Scientific American*, April 8, 2022, accessed September 4, 2023, https://www.scientificamerican.com/article/indias-inadvertent-missile-launch-underscores-the-risk-of-accidental-nuclear-warfare/

5 Debak Das, "Not Much Happened after India's Accidental Cruise Missile Launch into Pakistan—This Time," *Bulletin of the Atomic Scientists*, March 25, 2022, accessed September 4, 2023, https://thebulletin.org/2022/03/not-much-happened-after-indias-accidental-cruise-missile-launch-into-pakistan-this-time/

Debak Das, "Not Much Happened after India's Accidental Cruise Missile Launch into Pakistan-This Time," *CISAC*, March 25, 2022, accessed September 4, 2023, https://cisac.fsi.stanford.edu/news/not-much-happened-after-india%E2%80%99s-accidental-cruise-missile-launch-pakistan%E2%80%94-time

6 Ali Ahmed, "India-Pakistan: A Missile Misfired Opens Up Opportunity," *South Asian Voices,* March 25, 2022, accessed September 5, 2023, https://southasian-voices.org/india-pakistan-a-missile-misfired-opens-up-opportunity/

7 Daryl Kimball, "India Accidentally Fires Missile Into Pakistan," *Arms Control Association,* April 2022, accessed September 5, 2023, https://www.armscon-trol.org/act/2022-04/news/india-accidentally-fires-missile-into-pakistan#:~:text=India%20officially%20acknowledged%20it%20had,the%20two%20nuclear-armed%20neighbors

8 Amina Afzal, "The Luck Factor: Preventing Nuclear War in South Asia," *South Asian Voices,* March 25, 2022, accessed September 7, 2023, https://southasian-voices.org/the-luck-factor-preventing-nuclear-war-in-south-asia/

Michael Kugelman, "The Missile Crisis That Wasn't," *Foreign Policy*, March 17, 2022, accessed September 7, 2023, https://foreignpolicy.com/2022/03/17/india-pakistan-missile-launch-accidental-crisis-security/

9 Hassan Aslam Shad, "Did India Purposely Fire a BrahMos Missile at Pakistan?" *The National Interest,* March 21, 2022, accessed September 7, 2023, https://nationalinterest.org/blog/buzz/did-india-purposely-fire-brahmos-missile-paki-stan-201341

10 Manuel Herrera, "The Mian Channu Incident and the Enduring Risk of Nuclear Escalation between India and Pakistan," *IAI Istituto Affari Internazionali,* April 4, 2022, accessed September 7, 2023, https://www.iai.it/en/pubblicazioni/mian-channu-incident-and-enduring-risk-nuclear-escalation-between-india-and-pakistan

11 Anum Khan, "Indian Missile Crisis: One Step from Nuclear War?" *Asia-Pacific Leadership Network,* June 28, 2022, accessed September 9, 2023, https://www.apln.network/analysis/commentaries/indian-missile-crisis-one-step-from-nuclear-war

12 USIP, "Senior Study Group on Strategic Stability in Southern Asia," *United States Institute of Peace,* May 2022, accessed September 9, 2023, https://www.usip.org/programs/senior-study-group-strategic-stability-southern-asia

13 Ayaz Gul, "India Confirms It Accidentally Fired Missile into Pakistan," *Voice of America,* March 11, 2022, accessed September 11, 2023, https://www.voanews.com/a/india-confirms-it-accidentally-fired-missile-into-pakistan-/6480872.html
14 Michael Kugelman, "The Missile Crisis That Wasn't," *Foreign Policy,* March 17, 2022, accessed September 11, 2023, https://foreignpolicy.com/2022/03/17/india-pakistan-missile-launch-accidental-crisis-security/
15 Prasanna Zore, "'Why Would India Deliberately Fire an Unarmed Missile into Pakistan?'" *Rediff,* March 15 2022, accessed September 11, 2023, https://www.rediff.com/news/interview/dr-yogesh-joshi-why-would-india-deliberately-fire-an-unarmed-missile-into-pakistan/20220315.htm
16 Sudhi Ranjan Sen and Faseeh Mangi, "Errant Indian Missile Nearly Led to Pakistan Retaliatory Strike," *Bloomberg.com,* March 15, 2022, accessed September 11, 2023, https://www.bloomberg.com/news/articles/2022-03-15/errant-indian-missile-nearly-led-to-pakistan-retaliatory-strike
17 Christopher Clary, "The Curious Case of the Accidental Indian Missile Launch," *War on the Rocks,* March 17, 2022, accessed September 11, 2023, https://warontherocks.com/2022/03/the-curious-case-of-the-accidental-indian-missile-launch/
18 Department of Political Science. "Sumit Ganguly," accessed September 12, 2023, https://polisci.indiana.edu/about/faculty/ganguly-sumit.html
19 Sumit Ganguly, "Opinion: Two Steps India and Pakistan Can Take after Accidental Missile Launch," *South China Morning Post,* March 17, 2022, accessed September 12, 2023, https://www.scmp.com/comment/opinion/article/3170558/two-steps-india-and-pakistan-can-take-after-accidental-missile
20 ANI, "Husain Haqqani Gets Passport after Being Forced to Give up Pakistani Citizenship," *The Indian Express,* August 27, 2016, accessed September 12, 2023, https://indianexpress.com/article/world/world-news/husain-haqqani-gets-passport-after-being-forced-to-give-up-pakistani-citizenship-2999144/
21 Husain Haqqani, "What an Indian Missile Malfunction Says about US-Pakistan Relations," *The Hill,* March 18, 2022, accessed, September 13, 2023, https://thehill.com/opinion/international/598667-what-an-indian-missile-malfunction-says-about-us-pakistan-relations/
Husain Haqqani, "Indian Missile Must Bring Self-Reflection to Pakistan. Imran Khan Should Keep EU, US Close," *ThePrint*, March 19, 2022, accessed September 13, 2023, https://theprint.in/opinion/indian-missile-must-bring-self-reflection-to-pakistan-imran-khan-should-keep-eu-us-close/879469/
22 Hassan Aslam Shad, "Did India Purposely Fire a BrahMos Missile at Pakistan?" *The National Interest*, March 21, 2022, accessed September 13, 2023, https://nationalinterest.org/blog/buzz/did-india-purposely-fire-brahmos-missile-pakistan-201341
23 "India's 'Accidental' Missile Launch – Jinnah Institute," March 25 2022, accessed September 13, 2023, https://jinnah-institute.org/feature/indias-accidental-missile-launch/
24 Debak Das, "Stanford Freeman Spogli Institute for International Studies," accessed September 13, 2023, https://fsi.stanford.edu/people/debak-das
25 Debak Das, "Not Much Happened after India's Accidental Cruise Missile Launch into Pakistan—This Time," *Bulletin of the Atomic Scientists,* March 25, 2022, accessed September 13, 2023, https://thebulletin.org/2022/03/not-much-happened-after-indias-accidental-cruise-missile-launch-into-pakistan-this-time/
26 SAV staff, "The Crisis That Wasn't: A Review of India's Accidental Missile Launch into Pakistan," *South Asian Voices-Stimson*, March 25, 2022, accessed

September 15, 2023, https://southasianvoices.org/the-crisis-that-wasnt-a-review-of-indias-accidental-missile-launch-into-pakistan/

27 Frank O'Donnell, "SAV Q&A with Frank O'Donnell: Key Takeaway's from India's Missile Malfunction," *South Asian Voices,* March 25, 2022, accessed September 15, 2023, https://southasianvoices.org/sav-qa-with-frank-odonnell-key-takeaways-from-indias-missile-malfunction/

28 *South Asian Voices.* "Ali Ahmed, Author at South Asian Voices," n.d, accessed September 15, 2023, https://southasianvoices.org/author/aliahmed/

29 Ali Ahmed, "India-Pakistan: A Missile Misfired Opens Up Opportunity," *South Asian Voices*, March 25, 2022, accessed September 15, 2023, https://southasianvoices.org/india-pakistan-a-missile-misfired-opens-up-opportunity/

30 Stimson Center, "Amina Afzal • Stimson Center," accessed September 15, 2023, https://www.stimson.org/ppl/amina-afzal/

31 SAV staff, "The Crisis That Wasn't: A Review of India's Accidental Missile Launch into Pakistan," *South Asian Voices-Stimson*, March 25, 2022, accessed September 16, 2023, https://southasianvoices.org/the-crisis-that-wasnt-a-review-of-indias-accidental-missile-launch-into-pakistan/

32 Amina Afzal, "The Luck Factor: Preventing Nuclear War in South Asia," *South Asian Voices*, March 25, 2022, accessed September 16, 2023, https://southasianvoices.org/the-luck-factor-preventing-nuclear-war-in-south-asia/

33 "Daryl G. Kimball, Executive Director | Arms Control Association," accessed September 16, 2023, https://www.armscontrol.org/about/daryl_kimball

34 Daryl Kimball, "India Accidentally Fires Missile Into Pakistan," *Arms Control Association,* April 2022, accessed September 16, 2023, https://www.armscontrol.org/act/2022-04/news/india-accidentally-fires-missile-into-pakistan#:~:text=India%20officially%20acknowledged%20it%20had,the%20two%20nuclear-armed%20neighbors

35 Khokar Riaz and Asma Khalid. "The Indian Missile Launch in Pakistan: A Skeptical View," *Atlantic Council,* April 5, 2022, accessed September 17, 2023, https://www.atlanticcouncil.org/blogs/southasiasource/the-indian-missile-launch-in-pakistan-a-skeptical-view/

36 Khokar Riaz and Asma Khalid. "The Indian Missile Launch in Pakistan: A Skeptical View," *Atlantic Council,* April 5, 2022, accessed September 17, 2023, https://www.atlanticcouncil.org/blogs/southasiasource/the-indian-missile-launch-in-pakistan-a-skeptical-view/

37 Adil Sultan, "India's Missile Launch: Was It Accidental or Deliberate?" *Strafasia,* April 7, 2022, accessed September 17, 2023, https://strafasia.com/indias-missile-launch-was-it-accidental-or-deliberate/

38 Princeton, "Zia Mian | Princeton Science & Global Security," n.d, accessed September 17, 2023, https://sgs.princeton.edu/team/zia-mian

39 M. V. Ramana and Zia Mian, "India's Inadvertent Missile Launch Underscores the Risk of Accidental Nuclear Warfare," *Scientific American,* April 8 2022, accessed September 18, 2023, https://www.scientificamerican.com/article/indias-inadvertent-missile-launch-underscores-the-risk-of-accidental-nuclear-warfare/

40 "About Us," *China Military Online,* accessed September 18, 2023, http://eng.chinamil.com.cn/SinglePage/AboutUs/index.html

41 Fang Xiaozhi, "Behind India's Missile 'Misfiring' Incident," *China Military Online,* April 20 2022, accessed September 18, 2023, http://eng.chinamil.com.cn/OPINIONS_209196/Opinions_209197/10149320.html

42 USIP, "USIP Senior Study Group Final Report- Enhancing Strategic Stability in Southern Asia," May 17, 2022, https://cis.mit.edu/sites/default/files/documents/enhancing-strategic-stability-in-southern-asia-final-report.pdf .pp 13–19

43 Lalwani Sameer, "How to Deal with the Risk of Nuclear Escalation in South Asia," *United States Institute of Peace*, May 17, 2022, accessed September 20, 2023, https://www.usip.org/publications/2022/05/how-deal-risk-nuclear-escalation-south-asia

44 Daniel Markey, "China, India and Pakistan: Tenuous Stability Risks Nuclear War," *United States Institute of Peace*, May 17 2022, accessed September 20, 2023, https://www.usip.org/publications/2022/05/china-india-and-pakistan-tenuous-stability-risks-nuclear-war

45 Daniel Markey, "Daniel Markey on Strategic Stability in Southern Asia," *United States Institute of Peace*, May 19 2022, accessed September 20, 2023, https://www.usip.org/publications/2022/05/daniel-markey-strategic-stability-southern-asia

46 Anum Khan, "Indian Missile Crisis: One Step from Nuclear War? | Asia-Pacific Leadership Network," 28 June 2022, accessed September 21, 2023, https://www.apln.network/analysis/commentaries/indian-missile-crisis-one-step-from-nuclear-war

47 Aarish Khan, "Reassessing the BrahMos Missile That Landed in Pakistan," September 29 2022, accessed September 21, 2023, https://thediplomat.com/2022/09/reassessing-the-brahmos-missile-that-landed-in-pakistan/

48 CSDR, "Skirting the Nuclear Threshold: Conflict De-Escalation and Strategic Stability in South Asia," *CSDR New Delhi,* March 31 2022, accessed September 21, 2023, https://www.youtube.com/watch?v=65h66CFQv1w

49 Feroz Hassan Khan, "Strategic Risk Management in Southern Asia," *Journal for Peace and Nuclear Disarmament* 5, no. 2 (August 18, 2022): 369–93, https://doi.org/10.1080/25751654.2022.2136878

50 Daniel Markey, "The Persistent Threat of Nuclear Crises Among China, India and Pakistan," United *States Institute of Peace,* February 1 2023, accessed September 23, 2023, https://www.usip.org/publications/2023/02/persistent-threat-nuclear-crises-among-china-india-and-pakistan

51 Debak Das, "The State of Nuclear Instability in South Asia: India, Pakistan, and China," *Lawfare*, September 3, 2023, accessed September 23, 2023, https://www.lawfaremedia.org/article/the-state-of-nuclear-instability-in-south-asia-india-pakistan-and-china

52 Manuel Herrera, "The Mian Channu Incident and the Enduring Risk of Nuclear Escalation between India and Pakistan," IAI *Istituto Affari Internazionali*, April 4, 2022, accessed September 17, 2023, https://www.iai.it/en/pubblicazioni/mian-channu-incident-and-enduring-risk-nuclear-escalation-between-india-and-pakistan

53 Manuel Herrera, "The Mian Channu Incident and the Enduring Risk of Nuclear Escalation between India and Pakistan," IAI *Istituto Affari Internazionali*, April 4, 2022, accessed September 17, 2023, https://www.iai.it/en/pubblicazioni/mian-channu-incident-and-enduring-risk-nuclear-escalation-between-india-and-pakistan

CONCLUSION

This book is a fine-grained history which uses a constructivist approach and undertakes an in-depth discourse analysis of the Indian, Pakistani and international discourses to understand the accidental launch of a supersonic cruise missile by one nuclear weapons state into the territory of another at a time of relative calm in a relationship otherwise characterized by hostility, tension and deep mistrust. The chapters in the book look at the role of the two governments, media communities, strategic communities, global media and international experts. Interesting insights emerge from this sustained examination of the events, actors and institutions on the Indian side, Pakistani side and at the international level.

The analysis of the official reactions of the two governments (and their anonymous statements to the press) in Chapter 1 helps to understand the evolution of the stated rationales for the accident from the Indian defense establishment's initial claim of the accident being a result of a technical malfunction to the eventual admission by the Court of Inquiry that human error was the primary cause. While a technical defect in the missile has implications for the reliability of the BrahMos as a "frontline weapon" for the Army, Navy and Air Force,[1] the "human error" angle points to potential deficiencies in personnel training and in the design of standard operating procedures (SOPs) for the handling of delivery platforms. Both chains of causality, unless redressed, have important ramifications for South Asian security. The lone article in Chapter 1 informing about the designers of the BrahMos missile being upset by the Indian government's initial statement blaming the accident on a technical malfunction raises (but does not conclusively resolve) the tantalizing question of whether internal dissension within

DOI: 10.4324/9781003543909-7

the Indian strategic establishment had any role to play in the transition from the technical malfunction to the human error rationale.

Given the sensitivity of issues surrounding the missile accident, both governments used anonymous leaks to the media to convey information that they deemed important although such statements were more numerous on the Indian side and seemed intended to achieve a slow trickling out of information without official attribution. The end-goal of such disclosures may have been to convey certain important details (such as the missile being unarmed and being accidentally released during a practice exercise) to Pakistan or to sections of the strategic community in India or both. While anonymous defense sources on the Indian side have claimed that the fallout from the accident was contained by an "immediate intimation by India (to Pakistan) over a military hotline,"[2] such claims are at variance with Pakistani complaints that India did not use the hotline.[3]

Also striking is the continued ambiguity on important issues. Where was the missile launched from? While Pakistani official accounts claim that the missile was launched from Sirsa, the Indian official account makes no mention of the launch site, with anonymous official statements to the media claiming that the missile was launched from Ambala. Although there is no known BrahMos base near Sirsa and there is one near Ambala,[4] this does not entirely obviate the possibility that the missile could have been accidentally launched from a secret base in Sirsa.

Regarding whether the BrahMos is nuclear-capable[5] or not[6], multiple articles in the Indian press have claimed both. Despite the US classification of the BrahMos as a conventional missile, the Pakistani government continues to view it as nuclear-capable as is clear from the statement of its foreign minister. This ambiguity regarding whether the BrahMos is conventional or dual use can also be observed in the scholarly discourse.[7]

The official Pakistani perception of the BrahMos as nuclear-capable has important implications in terms of its threat perception from the BrahMos and the retaliatory options it would deploy in the event of a conflict. There are also claims by Indian media articles (quoting anonymous government sources) of both the presence and absence of a self-destruct mechanism on the BrahMos. Given that one of the seven questions posed by Pakistan to India was precisely about the presence or absence of this self-destruct capability, the available evidence from the documentary record on the Indian side during the period under examination does not point to a definitive conclusion.

Chapter 2's focus on the media coverage of the missile accident provides interesting insights including the framing strategies used by the Indian and Pakistani media. Some sections of the Indian media displayed considerable objectivity by praising Pakistan for its restrained response and by constructively criticizing the Indian government for the accident and for the delay in its acknowledgement of it. However, this criticism was carefully calibrated to support the primacy of the Indian Court of Inquiry and to not endorse exploration of courses of action such as a joint probe with Pakistan.

Other more government-friendly media outlets deployed a variety of framing strategies to defend the Indian government including trying to minimize the accident by referring to the BrahMos as a tactical system, highlighting humor about the incident on Indian social media, giving coverage to conspiracy theorists in Pakistan, blaming China for instigating the Pakistani demand for a joint probe and claiming that the Indian CoI's human error conclusion had belied the Pakistani claim of a deliberate Indian launch. More hawkish elements in the Indian media, albeit a minority, embraced Pakistan's allegations of the missile launch being a deliberate ploy and triumphantly claimed that it had exposed weaknesses in Pakistan air defense systems.

For its part, the Pakistani media's framing strategy attempted to undermine India's claims of being a responsible state by highlighting India's admission of its mistake, Pakistan's restrained response (despite initially considering retaliation) and praise for such restraint by Indian journalists. The Pakistani media also attempted to connect the missile's accidental launch to other safety-related lapses to question India's overall safety and security culture as a critical shortcoming that was being hidden by the Indian Court of Inquiry and its inadequate final conclusion. The Pakistani's media lack of trust in India was also evident in its speculation that India could have deliberately launched the missile to test Pakistani defenses. Alongside this mistrust also lay a certain confidence that Pakistan would now be able to reverse engineer the crashed missile.

Chapter 3 dealt with the reactions of the Indian and Pakistani strategic communities to the missile accident. Reactions in the Indian strategic community included a section of strategists commending Pakistan for its mature and restrained response and a smaller, nationalistic, group exulting that the accidental launch was a deliberate (and successful) Indian attempt to test Pakistani air defenses.

On the other hand, a significant section of the Pakistani strategists analyzed India's non-use of the military hotline, non-use of the purported self-destruct option on the cruise missile and silence in response to the DG-ISPR conference as part of a deliberate Indian strategy of needling Pakistan to test its military readiness. Especially salient is the view of General Khalid Kidwai, adviser to Pakistan's National Command Authority and Director General of the Strategic Plans Division regarding the BrahMosmissile intrusion being a result of the rise of Hindutva fundamentalism in India and the aggressive political rhetoric, increased risk-taking and capability development (counterforce capabilities) resulting from it. Also remarkable is Kidwai's speculation that the missile intrusion was India's attempt towards establishing a pattern of such intrusions into Pakistan that can be normalized. The aforementioned Pakistani perception of the missile intrusion as a deliberate ploy by a fundamentalist Indian government has serious implications for Pakistani threat perception and response in case another entirely accidental missile intrusion takes place during a period of tension in the overall relationship.

The chapter also revealed less discussed factors that could have played a role in the missile accident such as the supposedly tiered structure of the Indian Air Force in which first rung officers are appointed as pilots while second rung officers are assigned to missile units with implications for performance of such units. The key insight from the chapter for India-Pakistan relations is that underlying issues such as the deep mistrust, lack of doctrinal clarity, the non-utilization of military and foreign secretary level hotlines and lack of top-level political communication continue to affect India-Pakistan relations. In such an atmosphere of low information and low trust, the risk always exists of an exaggerated over-interpretation of one country's actions by another, resulting in the ascribing of deliberate motives to purely accidental events.

Chapter 4 focused on the role of the international media in covering the missile accident. Overall, the international media's coverage of the accidental launch was scant with only a few articles and news segments, especially given the seriousness of the incident as perhaps the first accidental launch of a supersonic missile by one nuclear power into the territory of another.[8] Factors such as the incident not resulting in any casualties, the pre-occupation with the war in Ukraine, the overreliance on news agencies like *Reuters*, the brevity of Indian official statements and the secrecy surrounding the Indian Court of Inquiry were responsible for the sporadic coverage.

Even so, certain revelations made by the international media (such as Pakistan's claimed plans of retaliatory strike or the claims by anonymous Indian experts of the Indian move to shutdown missile systems after the launch) were important in either adding strategically crucial insights to the discourse or in adding more credibility to existing revelations made by the Indian and Pakistani media. Certain international media outlets thus became part of the attempts by both the Indian and Pakistani security establishments to shape the narrative surrounding the accident. Missing from the global media's focus on the immediate developments surrounding the accident and its attempts to contextualize the misfire in terms of the previous outbreaks of war between India and Pakistan was a deeper analysis of the incident's implications for South Asian security.

Finally, Chapter 5 focused on the analyses of the missile accident advanced by international experts in the US, China and other countries. Several important technical and strategic issues were discussed including the circumstances that led to the unexpected launch of the missile during routine maintenance, shortcomings in the design of the BrahMos missile system, the standard operating procedures between India and Pakistan to deal with accidents involving advanced weapons systems such as cruise missiles and whether India's offensive strike missiles were deployed in launch-ready conditions.

Experts also discussed the implications of the missile accident for India's attempt to continually de-hyphenate itself from Pakistan as a responsible nuclear state, India's command and control systems and the export-related commercial incentives that may have led India to blame the accident on human error rather than the technical malfunction explanation proffered earlier by the Indian establishment which would have compromised exports.

Speculations were also made regarding the missile being launched either due to an external cyberattack or by insiders with covert political authorization. Recognizing the role of good fortune in preventing escalation, various recommendations were also made regarding the need for more robust and reliable mechanisms of crisis communication, regular dialogue between political and military establishments, need to bring cruise missiles under the ambit of the 2005 ballistic missile pre-notification agreement and re-programming of guidance systems for missiles with co-ordinates for unoccupied regions such as the Thar desert.

Overall, a constructivist analysis of the domestic discourse in India and Pakistan and of the international discourse on the missile accident of March 2022 holds implications for state policy in South Asia on a number of issues. Given the seriousness of the accident (despite it not causing any casualties) and the potential for serious escalation if it does result in casualties or damage to a strategic facility, there is an urgent need for a serious bilateral (inter-state) and domestic (intra-state) conversations on communication by the governments in the aftermath of the crisis.

At the bilateral level, discussions at the political and the military levels need to focus on expanding the scope of the 2005 Agreement on the Pre-notification of Ballistic Missile Tests to also include cruise missiles. The Agreement requires both governments to notify each other "no less than three days in advance" of a flight test of "any land or sea launched, surface-to-surface ballistic missiles." The pre-notification is required to be conveyed through the "respective Foreign Offices and High Commissions" in a specific format specified in the Agreement. Further, India and Pakistan are also required to ensure that the test launch site and the planned impact area do not fall within 40 km and 70 km respectively, of the International Boundary (IB) or the Line of Control (LOC) on the side of the country testing the ballistic missile. The Agreement is clear that the ballistic missile being tested shall not cross the IB or the LOC and shall "maintain a horizontal distance of at least 40 kms" from the IB and the LOC.[9]

Although the dynamics of a cruise missile accident are likely to be significantly different from the test of a ballistic missile, many provisions of the aforementioned agreement such as the specific format required for notification can be adapted to require the government from whose territory the accidental launch has occurred to immediately notify the other government regarding its time of occurrence and time of usage of the military hotline. A similar adaptation of the distance requirements with respect to the test launch site and the planned impact area as also horizontal distance requirements from the IB and LOC can be discussed keeping in view the particularities of cruise missiles.

At the domestic level, the provision of timely and accurate information through official statements regarding the details of the accident and the use of the military hotline is crucial for accurate reportage in the domestic (and international) media and for a constructive debate in the Indian, Pakistani and international strategic communities. While revealing the details of the missile that suffered the accident and the location from which it was launched may be deemed as too sensitive on national security grounds by either India or Pakistan, clarity regarding the causal factors of the accident (to the extent possible) and the use of the military hotline can certainly be expected. Official silence in the aftermath of the accident, terse statements that occlude crucial details and anonymous leaks to the media on some selective aspects of the accident as part of a narrative-shaping exercise run the risk of encouraging rampant speculation in the media and strategic communities skewing the discourse (especially in Pakistan) towards a threat perception and policy recommendations that may not be in alignment with the real nature of the threat.

At a broader level, there is also an acute need for the resumption of dialogues at the Track 2 level between the respective political and military establishments on issues pertaining to nuclear doctrine and doctrinal clarity, prevention of entanglement of conventional and strategic weapons systems and a threadbare discussion on the advertence potential of emerging technologies such as UAVs and AI. The aforementioned policy recommendations serve as a reiteration of similar suggestions that have been made by South Asian security experts in the years since the 1998 tests.

Even as the aforementioned inter-state dialogues are attempted, both governments should also initiate serious intra-government, inter-agency consultations on the existing command and control architecture, current standard operating procedures (especially regarding inspection), target-setting, launch authorization and launch sequence.

Litigation over the Accident

Despite the Indian government's attempts at bringing the proceedings surrounding the missile accident to a close with the announced dismissal of an unidentified wing commander, a group captain and a squadron leader[10] on 22 August 2022, the incident continues to linger in the attention space of Indian policy and media elites. The identity of one of three dismissed officers was revealed on 15 March 2023 when Wing Commander Abhinav Sharma filed a case against India's Defense Ministry and the Chief of the Indian Air Force (IAF) in the Delhi High Court claiming wrongful dismissal.[11] While the Court of Inquiry's findings were not made public, press reports claimed that Sharma and the two other IAF officers had been dismissed by the government on 22 August 2022 after they were found guilty for "failing to ensure the disconnection of combat connectors of all missiles loaded on Mobile Autonomous Launcher (MAL) prior to convoy movement, thus violating the combat SOP (Standard Operating Procedure)."[12]

The petition filed by Sharma and carried in the media also confirmed that Sharma was accused of "failure to perform external checks of the combat connectors of combat missiles, which continued to remain connected to FCS Junction Box 1 & 3"[13] and for "failing to intervene to prevent the MAL from engaging in an unsafe act of launching the missile from the station, which resulted in the accidental firing."[14] Thus, Sharma and the two other officers were found guilty of "gross negligence" which led to the firing.[15]

> The Delhi High Court on Tuesday issued notice to the Defence Ministry and others on a petition moved by an Air Force officer, who was terminated after the accidental launch of BrahMos cruise missile that landed in Pakistan in March last year. A division bench of Justices Suresh Kumar Kait and Neena Bansal Krishna directed the Ministry of Defence, Chief of Air Staff and others to respond to former Wing Commander Abhinav Sharma's plea within six weeks. The sacked officer was granted four weeks to file his rejoinder thereafter.…As per the petitioner, he was posted as an engineering officer to the squadron and he was imparting professional and practical training only for duties that were pure of a maintenance nature. It was also argued that he had no experience in conducting operations and handling operational emergencies. "The petitioner was not trained against the counts of blame apportioned to him in the court of inquiry and he acted in complete obedience to the SOP. The petitioner had no experience in conducting operations and handling operational emergencies and the respondents acted in a completely malafide manner by issuing the termination order," his plea read.The petitioner also argued that he was denied the opportunity to stand trial by a court martial as the authorities invoked Section 18 of the Air Force Act.[16]

Sharma submitted before the court that the missile accident was not just due to an operational failure resulting from a deviation from SOPs but a broader policy failure on the part of Squadron leader, Air Commodore and the Air-Vice Marshal.

> Sharma has alleged that the policy failure which led to the misfiring is being camouflaged by terminating him under Section 18 of the Air Force Act. He has further accused the Air Vice Marshal (Assistant Chief of the Air Staff (operations) of absolving himself and giving a clean chit to the Squadron leader and the Air Commodore without carrying out a proper inquiry. Sharma has accused the Air Commodore and Squadron Leader of ignoring and overriding the safety precaution popups and ignoring the bright and conspicuous Red colour warning regarding the danger of firing a Live Weapon, the report said. The Wing Commander, in his petition filed in the Delhi High Court last month, has alleged that the Squadron leader ordered an "alternate mission sequence" for "reasons best known to him" and informed the Air Commodore "of the change on the secured operations communication network, which bypassed the safety check of utilising live weapons during simulated exercise", the *ET report* said quoting his petition in court. The petition, further said that the said communication occurred was available only to the Squadron Leader, and the Air Commodore. And adds that the Air Vice Marshal did not even pose a question or seek an explanation from the Air Commodore and Squadron Leader while presiding over the inquiry court. According to Sharma, before the accidental firing of the missile the Squadron had conducted training exercises using "Live" weapons on several occasions by adopting certain processes. Sharma has alleged that the Air Force officials were aware of the inherent risks of using Live Weapons for training purposes. The Directorate of Air Staff Inspection at Air Headquarters (DASI Team) had flagged the risks in the use of "Live" weapon deployment for conducting planned simulation exercises in January 2022.[17]

More specifically, Sharma alleged that existing SOPs did not include any prohibition on the use of live weapons during simulated exercises and that such a prohibition was only issued by the Assistant Chief of the Air Staff (Operations) after the missile accident. Such a revision of SOPs was clear proof that the missile accident was a policy failure which was being suppressed by scapegoating Major Sharma under Section 18 of the Air Force Act.

> The Petitioner performed all his duties as per the SOP dated 15.03.2021. A perusal of the said SOP, which governed the operations on the fateful day would demonstrate that the Impugned Termination Order is malafide and the counts of blame levelled against the Petitioner bear no nexus with his responsibility as per his role prescribed by the SOP. The cause of

> the unfortunate incident was solely operational in nature," the petition said....The petition said on the date of the unfortunate incident, there was no specific regulation or prohibition on the use of live weapons during simulated exercises despite the authorities being aware of the inherent risks associated with such processes and the revised SOP was issued only after the accident. "Immediately thereafter Respondent No. 5 [Assistant Chief of the Air Staff (Operations)] caused issuance of a revised SOP dated 25.04.2022 which removed the ambiguities and deficiencies in the SOP dated 15.03.2021 with regard to the use of 'Live' weapons during training exercises. The same clearly demonstrates that the accidental firing of the missile was a policy failure which is being camouflaged by invoking Section 18 of the Air Force Act and terminating the petitioner," it alleged. "Respondent 5, localized liability to the Squadron level in order to camouflage the Policy failure at the HQ level"....it added.[18]

The Indian government, in its response to Wing Commander Sharma's petition contended before the court that the accidental fire of the missile "cost India Rs 24 crore,"[19], could have led to a warlike situation and caused major embarrassment before the international community.[20] The government also explained to the court the reasons for Sharma's dismissal under Section 18 of the Air Force Act as a decision to initiate action against the three erring officers under Section 19 of the Air Force Act 1950 and Rule 16 of the Air Force Rules 1969 "by issuing a Show Cause Notice for dismissal/removal from the Service would bring the sensitive and secret issues in full public domain which would be prejudicial to the security interests of the State"[21]

> The accidental firing of a BrahMos combat missile into Pakistan in March last year affected the relations of the nation with the neighbouring country and led to a loss of ₹24 crore to the state exchequer, the Union government told the Delhi high court while justifying the dismissal of three IAF (Indian Air Force) officers, including a Wing Commander, for gross negligence. Opposing a petition by Wing Commander Abhinav Sharma against his termination from service, the Centre in a short affidavit said the trial of three officers by a court martial was "inexpedient", especially considering the sensitive nature of evidence on record and "also the fact that the international community was interested to know the important practical details regarding the firing of missile". "Considering the sensitive nature of the subject matter having widespread ramifications for the security of the State, a conscious and considered decision was taken in good faith to terminate the service of the petitioner under the President's pleasure clause. Such a decision has been taken in the Indian Air Force after 23 years as facts and circumstances of the case warranted such action," the Union government said.[22]

Arguing for the government, Additional Solicitor General (ASG) Sharma pointed out to the Court that the "petitioner is gainfully employed with a multinational company with a heavy salary" and had only approached the court after more than six months following his termination.[23]

Wing Commander Sharma's petition before the court and the details contained therein raise interesting questions regarding specific technical and organizational aspects pertaining to the missile accident.

Does the BrahMos have an alternative mission sequence that can be ordered by the Squadron leader, allowing for the overriding of safety precautions before it is launched? Does the implementation of such a firing sequence only require communication between the Squadron leader and the Air Commodore over a secure operations communications network? Did Sharma's squadron conduct multiple exercises with live weapons before the missile accident? Were IAF officials indeed aware of the risks of using live weapons for training? And did the Directorate of Air Staff Inspection at Air Headquarters (DASI Team) indeed flag the risk of using live (albeit unarmed) weapons to conduct simulation exercises in January 2022, a full two months before the accident? Given the serious implications of such details about the missile accident for national security, the answers to the aforementioned questions may perhaps never be known.

More recently, the IAF has responded to Wing Commander Sharma's petitition revealing the identities of all three dismissed officers, number of witnesses examined during the Court of Inquiry, and the identity of the Air Commodore whom Sharma had blamed for causing the accident.

> According to a report by The Economic Times, the IAF told the High Court that combat connectors of the BrahMos missile "remained connected to the junction box" which led to accidental firing in Pakistan....The IAF's Court of Inquiry examined 16 witnesses and found Group Captain Saurabh Gupta, Squadron Leader Pranjal Singh, and Wing Commander Abhinav Sharma – all members of the Combat Team – responsible for the incident. The incident caused "potential threat to any airborne/ground object/personnel and also causing damage to the reputation of Indian Air Force and the nation at large.... IAF denied blame on Air Commodore JT Kurien. Instead, the IAF said that "all activities" took place in the presence of Sharma. "All activities had taken place inside the Mobile Autonomous Launcher in his (Sharma) presence and he was a witness to the actions being taken by his Ops officer," the IAF said.[24]

Even as litigation over the missile crisis plays out in court, the accident continues to have implications for India-Pakistan relations and regional security. Former US Secretary of State Mike Pompeo has claimed in his book *Never Give an Inch: Fighting for the America I Love* that India and Pakistan came close to a nuclear exchange on the night of 27–28 February 2019 in the aftermath of the Indian strike at Balakot with Pompeo being informed by his Indian counterpart Minister of External Affairs Sushma Swaraj that India was preparing an escalatory nuclear response to Pakistani preparations for a nuclear strike on India.[25] Pompeo claimed credit for working overnight with both the Indian and Pakistani leadership to reduce tensions.[26]

While commentary in India and Pakistan[27] has called for such "over-hyped" claims of US indispensability to India-Pakistan crisis mediation to be tempered with a more "refined understanding of deterrence stability in South Asia,"[28] a worrying question still arises regarding the impact of a missile accident occurring during a period of elevated India-Pakistan tensions such as in the days following the Pulwama attack and after the Balakot airstrike. American academic Christopher Clary has reckoned that the Balakot crisis "offers a template of circumstances in which an errant cruise-missile launch could have proved catastrophic rather than merely embarrassing."[29] Indian strategic affairs commentator Sushant Singh agrees that "sheer good luck" resulting from the missile being unarmed, not landing on a military facility, hitting an aircraft or claiming lives in a populated area resulted in the accident not causing a full-blown security crisis.[30]

Ultimately, the March 2022 missile accident is a warning of future security risks to South Asia from the dangerous nexus of antagonistic nationalisms, festering territorial disputes, intense security competition, deep distrust, growing numbers of nuclear weapons, perceived lack of doctrinal clarity, growing sophistication of missile arsenals, and growing complexity of their deployment postures.

Genuine political will to address territorial disputes, buy-in from the respective security establishments for risk reduction measures, utilization of existing crisis de-escalation mechanisms such as the Foreign Secretary and DGMO hotlines, expansion of existing agreements such as the 2005 ballistic missiles pre-notification agreement to also include cruise missiles and other measures as part of a proactive policymaking approach is imperative to avoid a future, sudden-onset, serious security crisis.

Notes

1 S. Jain, "BrahMos: Building World's 'Fastest Supersonic Cruise Missile', Future Variants & Export Orders," *The Times of India*, June 14, 2023, accessed October 2, 2023, https://timesofindia.indiatimes.com/business/india-business/brahmos-building-worlds-fastest-supersonic-cruise-missile-future-variants-export-orders/articleshow/100979055.cms?from=mdr

2 V. Thapar, "Commanding Officer, Others To Blame For Accidentally Firing BrahMos Missile Into Pakistan: Reports," *BW Businessworld*, April 11 2022, accessed October 11, 2023, http://businessworld.inhttps://www.businessworld.in/article/Commanding-Officer-Others-To-Blame-For-Accidentally-Firing-BrahMos-Missile-Into-Pakistan-Reports/11-04-2022-425201

S. A. Philip, "Missile Accidentally Fired 'during Inspection' at Secret IAF Base, Pakistan 'Didn't Track,'" *ThePrint*, March 12, 2022, accessed October 11, 2023, https://theprint.in/defence/missile-accidentally-fired-during-inspection-at-secret-iaf-base-pakistan-didnt-track/870814/

3 M. Korda, "Flying Under The Radar: A Missile Accident in South Asia," *Federation of American Scientists*, April 4 2022, accessed October 11, 2023, https://fas.tghp.co.uk/publication/flying-under-the-radar-a-missile-accident-in-south-asia/

4 M. Korda, "Flying Under The Radar: A Missile Accident in South Asia," *Federation of American Scientists*, April 4 2022, accessed October 11, 2023, https://fas.tghp.co.uk/publication/flying-under-the-radar-a-missile-accident-in-south-asia/

5 A. Sinha, "Misfire of BrahMos Missile Not Cause for Any Specific Concern: IAEA," *The Indian Express*, November 14, 2022, accessed October 12, 2023, https://indianexpress.com/article/world/misfire-of-brahmos-missile-not-cause-for-any-specific-concern-iaea-8266709/

6 Ajai Shukla, "Brahmos Can't Be Mistaken for Nuclear-Tipped Missile," *Business Standard*, March 15, 2022, accessed October 12 2023, https://www.business-standard.com/article/economy-policy/brahmos-can-t-be-mistaken-for-nuclear-tipped-missile-122031401493_1.html

7 Stimson, "India and Nuclear Asia: Forces, doctrine, and Danger," *The Stimson Center*, January 17, 2019, accessed October 13 2023, https://www.stimson.org/wp-content/files/EventTranscript_1_17_IndiaNuclearAsia.pdf .p5
Yogesh Joshi, Frank O' Donnell, and Harsh Pant, "India's Evolving Nuclear Force and Implications for U.S. Strategy in the Asia-Pacific," *USAWC Press*, June 1, 2016, accessed October 14 2023, https://press.armywarcollege.edu/cgi/viewcontent.cgi?article=1301&context=monographs .p3
Ms Tanzeela Khalil, "Trends in India's Nuclear Force Modernization: Regional and Global Implications" *CISS Insight Journal of Strategic Studies* 7, no. 2 (2019): 62–88, https://journal.ciss.org.pk/index.php/ciss-insight/article/view/13 .p 78
J. Mitra, "Nuclear BrahMos: On the Anvil? • Stimson Center," *Stimson Center*, September 18, 2020, accessed October 14 2023, https://www.stimson.org/2020/nuclear-brahmos-on-the-anvil/

8 C. Clary, "The Curious Case of the Accidental Indian Missile Launch," *War on the Rocks*, March 17, 2022, accessed October 15, 2023, https://warontherocks.com/2022/03/the-curious-case-of-the-accidental-indian-missile-launch/

9 W. Mechanix, "Agreement between India and Pakistan on Pre-notification of Flight Testing of Ballistic Missiles," *Stimson Center*, Ocotober 25, 2012, https://www.stimson.org/2012/agreement-between-india-and-pakistan-on-pre-notification-of-flight-tes/

10 B. Ajeyo, "IAF Officer Approaches Delhi High Court against Defence Ministry, Air Force Chief over BrahMos Incident," *Firstpost*, March 15, 2023, accessed October 15, 2023, https://www.firstpost.com/india/iaf-officer-approaches-delhi-high-court-against-defence-ministry-air-force-chief-over-brahmos-incident-12294822.html

11 K. Nakshab, "Brahmos Misfiring into Pakistan: Wing Commander Abhinav Sharma Files Case against Defence Ministry and IAF Chief," *Times Now*, March 14, 2023, accessed October 15, 2023, https://www.timesnownews.com/india/brahmos-misfiring-into-pakistan-wing-commander-abhinav-sharma-files-case-against-defence-ministry-and-iaf-chief-article-98638952
The Wire Staff, "Misfired Brahmos Which Landed in Pakistan Could Have Led to Warlike Situation: Union Govt to HC," *The Wire*, March 15, 2023, accessed

October 16, 2023, https://thewire.in/security/misfired-brahmos-which-landed-in-pakistan-could-have-led-to-warlike-situation-union-govt-to-hc
12 Nakshab K. TimesNow, "Brahmos Misfiring into Pakistan: Wing Commander Abhinav Sharma Files Case against Defence Ministry and IAF Chief," March 14, 2023, accessed October 16, 2023, https://www.timesnownews.com/india/brahmos-misfiring-into-pakistan-wing-commander-abhinav-sharma-files-case-against-defence-ministry-and-iaf-chief-article-98638952
13 PTI, "HC Seeks Centre's Stand on Challenge to Dismissal of Air Force Officer over Brahmos Misfire," *ThePrint*, March 14, 2023, accessed October 17, 2023, https://theprint.in/india/hc-seeks-centres-stand-on-challenge-to-dismissal-of-air-force-officer-over-brahmos-misfire/1443108/
14 Nakshab K. TimesNow, "Brahmos Misfiring into Pakistan: Wing Commander Abhinav Sharma Files Case against Defence Ministry and IAF Chief," March 14, 2023, accessed October 17, 2023, https://www.timesnownews.com/india/brahmos-misfiring-into-pakistan-wing-commander-abhinav-sharma-files-case-against-defence-ministry-and-iaf-chief-article-98638952
15 R. Banka, "Missile Misfire into Pak Could Have Led to a Warlike Situation: Govt to Delhi HC," *Hindustan Times* March 15, 2023, accessed October 18 2023, https://www.hindustantimes.com/india-news/missile-misfire-could-have-led-to-a-warlike-situation-govt-to-delhi-high-court-101678819003275.html
16 J. Wilson, "Accidental Launch of BrahMos Missile: Delhi HC Seeks Centre's Reply to Sacked Officer's Petition," *The New Indian Express,* March 15 2023, accessed October 18, 2023, https://www.newindianexpress.com/cities/delhi/2023/mar/15/accidentallaunch-of-brahmos-missile-delhi-hc-seeks-centres-reply-to-sacked-officers-petition-2556271.html
Express News Service, "BrahMos Missile Misfire: Delhi HC Seeks Centre's Reply on Ex-IAF Officer's Plea against Termination," *The Indian Express,* March 15, 2023, accessed October 20, 2023, https://indianexpress.com/article/india/brahmos-missile-misfire-hc-seeks-centres-reply-on-ex-iaf-officers-plea-against-termination-8497373/
17 BS Web, "Brahmos Accidental Firing a Case of Policy Failure: Sacked Wing Commander," *Business Standard*, April 3, 2023, accessed October 20, 2023, https://www.business-standard.com/india-news/brahmos-accidental-firing-a-case-of-policy-failure-sacked-wing-commander-123040300477_1.html
18 PTI, "HC Seeks Centre's Stand on Challenge to Dismissal of Air Force Officer over Brahmos Misfire," *ThePrint,* March 14, 2023, accessed October 21, 2023, https://theprint.in/india/hc-seeks-centres-stand-on-challenge-to-dismissal-of-air-force-officer-over-brahmos-misfire/1443108/
19 FP Staff, "'Accidental Firing of BrahMos Missile into Pakistan Cost India Rs 24 Crore,' Centre Tells Delhi High Court," *Firstpost,* May 30, 2023, accessed October 21, 2023, https://www.firstpost.com/india/accidental-firing-of-brahmos-missile-into-pakistan-cost-india-rs-24-crore-centre-tells-delhi-high-court-12668392.html
20 R. Banka, "Missile Misfire into Pak Could Have Led to a Warlike Situation: Govt to Delhi HC," *Hindustan Times* March 15, 2023, accessed October 22, 2023, https://www.hindustantimes.com/india-news/missile-misfire-could-have-led-to-a-warlike-situation-govt-to-delhi-high-court-101678819003275.html
21 R. Banka, "BrahMos Missile Misfiring into Pakistan Cost India ₹24 Crore, Centre Tells Delhi HC," *Hindustan Times*. May 30, 2023, accessed October 22, 2023, https://www.hindustantimes.com/india-news/india-justifies-dismissal-of-iaf-officers-over-brahmos-missile-firing-into-pakistan-citing-loss-of-rs-24-crore-to-exchequer-101685388238329.html

22 R. Banka, "BrahMos Missile Misfiring into Pakistan Cost India ₹24 Crore, Centre Tells Delhi HC," *Hindustan Times*. May 30, 2023, accessed October 22, 2023, https://www.hindustantimes.com/india-news/india-justifies-dismissal-of-iaf-officers-over-brahmos-missile-firing-into-pakistan-citing-loss-of-rs-24-crore-to-exchequer-101685388238329.html
The Wire Staff, "Misfired Brahmos Which Landed in Pakistan Could Have Led to Warlike Situation: Union Govt to HC," *The Wire,* March 15, 2023, accessed October 22, 2023, https://thewire.in/security/misfired-brahmos-which-landed-in-pakistan-could-have-led-to-warlike-situation-union-govt-to-hc
FP
Staff, "'Accidental Firing of BrahMos Missile into Pakistan Cost India Rs 24 Crore,' Centre Tells Delhi High Court," *Firstpost,* May 30, 2023, accessed October 23, 2023, https://www.firstpost.com/india/accidental-firing-of-brahmos-missile-into-pakistan-cost-india-rs-24-crore-centre-tells-delhi-high-court-12668392.html
R. Banka, "Missile Misfire into Pak Could Have Led to a Warlike Situation: Govt to Delhi HC," *Hindustan Times,* March 15, 2023, accessed July 23, 2023, https://www.hindustantimes.com/india-news/missile-misfire-could-have-led-to-a-warlike-situation-govt-to-delhi-high-court-101678819003275.html
R. Banka, "BrahMos Missile Misfiring into Pakistan Cost India ₹24 Crore, Centre Tells Delhi HC," *Hindustan Times,* May 30, 2023, accessed October 25, 2023, https://www.hindustantimes.com/india-news/india-justifies-dismissal-of-iaf-officers-over-brahmos-missile-firing-into-pakistan-citing-loss-of-rs-24-crore-to-exchequer-101685388238329.html

23 R. Banka, "Missile Misfire into Pak Could Have Led to a Warlike Situation: Govt to Delhi HC," *Hindustan Times,* March 15, 2023, accessed October 25, 2023, https://www.hindustantimes.com/india-news/missile-misfire-could-have-led-to-a-warlike-situation-govt-to-delhi-high-court-101678819003275.html

24 Livemint, "IAF Reveals What Led to Accidental Firing of BrahMos into Pakistan," *Mint,* March 30, 2024, https://www.livemint.com/news/india/iaf-reveals-what-led-to-accidental-firing-of-brahmos-into-pakistan-11711773409628.html

25 PTI, "Mike Pompeo Claims India Informed Him Pakistan Was Preparing for Nuclear Attack Post-Balakot Surgical Strike," *The Economic Times*, January 24, 2023, accessed October 26 2023, https://economictimes.indiatimes.com/news/international/world-news/mike-pompeo-claims-india-informed-him-pakistan-was-preparing-for-nuclear-attack-post-balakot-surgical-strike/articleshow/97288213.cms?from=mdr
PTI, "India Informed Me Pak Was Preparing for Nuclear Attack Post-Balakot Strike, Claims Pompeo," January 25 2023, accessed October 26, 2023, https://www.indiatoday.in/world/story/mike-pompeo-claims-india-informed-him-pak-was-preparing-for-nuclear-attack-post-balakot-surgical-strike-2326003-2023-01-25
PTI, "India, Pakistan Came Close to a Nuclear War, Claims Former U.S. Secretary of State in New Book," *The Hindu*, January 24, 2023, sec. India accessed October 27, 2023, https://www.thehindu.com/news/national/india-pakistan-came-close-to-a-nuclear-war-claims-former-us-secretary-of-state-in-new-book/article66429650.ece

26 Outlook Web Desk, "India Informed about Pakistan's Plan of Nuclear Attack after Balakot Surgical Strike: Former US Secretary of State Mike Pompeo," *Outlook,* January 25, 2023, accessed October 27, 2023, https://www.outlookindia.com/national/-india-informed-about-pakistan-s-plan-of-nuclear-attack-after-balakot-sirgical-strike-former-us-secretary-of-state-mike-pompeo-news-256516

27 A. Shah, "Pompeo, Swaraj and the Myth of Nuclear Escalation. *Strategic Vision Institute*," February 8, 2023, accessed October 27, 2023, https://thesvi.org/pompeo-swaraj-and-the-myth-of-nuclear-escalation-in-the-february-2019-india-pakistan-crisis/

28 M. Dhritiman, "Pope's Balakot Claim and the Need to Adopt a Nuanced Understanding of Indo-Pak Deterrence Relationship," *Vivekananda International Foundation,* February 3, 2023, accessed October 27, 2023, https://www.vifindia.org/article/2023/february/03/pompeos-balakot-claim-and-the-need-to-adopt-a-nuanced-understanding-of-indo-pak-deterrence-relationship

29 C. Clary, "The Curious Case of the Accidental Indian Missile Launch," *War on the Rocks,* March 17, 2022, accessed October 28, 2023, https://warontherocks.com/2022/03/the-curious-case-of-the-accidental-indian-missile-launch/
The Curious Case of the Accidental Indian Missile Launch - War on the Rocks

30 M. Mashal and M. Salman, "India Accidentally Fires a Missile at Pakistan. Calm Ensues," *The New York Times,* March 12, 2022, accessed October 28, 2023. World, https://www.nytimes.com/2022/03/12/world/asia/india-pakistan-missile.html

For Product Safety Concerns and Information please contact our EU representative GPSR@taylorandfrancis.com
Taylor & Francis Verlag GmbH, Kaufingerstraße 24, 80331 München, Germany

www.ingramcontent.com/pod-product-compliance
Lightning Source LLC
LaVergne TN
LVHW011818200926
848484LV00005B/1231

* 9 7 8 1 0 3 2 4 3 7 6 6 8 *